Contents

Introduction ....
Permutation Matrices 2
    Permutations: 2
    Permutation matrices: 3
    Aside: 4
    The identity permutation: 5
    Sequential combination of permutations: 6
    A note: 6
    The conventional notation - combining permutations: 7
    Aside: 8
    Multiplicative closure: 9
    Matrix multiplication: 9
    Inverse permutation matrices: 10
    The non-commutativity of permutation matrices: 12
    How many permutation matrices?: 12
    Summary: 13
    Addendum: 13
Determinants and Permutation Matrices 14
    Digression - the Levi-Civita symbol – notation only: 14
    End of digression - now to determinants: 15
    The matrix determinant: 15
    Back to the Levi-Civita symbol: 17
    The Levi-Civita symbol and permutation matrices: 18
    The determinants of permutation matrices: 19
    The matrix determinant written using permutation matrices: 20
    General properties of determinants: 21

- The nature of the determinant: 21
- Aside: 22

Odd and Even Permutation Matrices 24
- Even permutation matrices are a closed set: 24
- The inverse of an even permutation is an even permutation: 24
- Odd for even: 25

The Order of a Permutation Matrix 26
- Aside: 26
- The order of a permutation matrix: 27
- A few examples: 27
- Cyclic groups of permutation matrices: 28
- Cyclic groups: 29
- Symmetric groups: 30
- More than one generating element: 31
- Summary: 31

Finite groups and Representations 32
- Finite group definition: 32
- The order of a finite group: 33
- Seeking finite groups: 33
- The order 1 finite group: 34
- Representations of finite groups: 34
- The order 2 finite group: 34
- Aside: 35
- The order 3 finite group: 36
- The C_2 subgroups of S_3: 37
- Aside: 37
- Back to the order three group: 38

Contents

- Back to the symmetric groups: 38
- Groups other than the symmetric groups: 39
- All finite groups are subgroups of symmetric groups: 40
- Symmetric groups as subgroups: 40
- Summary: 42

Special Representations of Finite Groups 43
- The fundamental representation of a group: 43
- The adjoint representation of a group: 44
- A peek ahead: 45
- Adjoint representations and Cayley tables: 46
- Summing permutation matrices: 48
- Warning: 49

Casting out the Order Four Groups 50
- Aside: The order five groups: 53
- Summary: 54
- Repeated warning: 54

Adjoint Representations and Cyclic Groups 55
- Cyclic groups: 56
- Even and odd permutations and the cyclic groups: 57
- Aside: 58
- The adjoint cyclic group: 58

Cosets 59
- Cosets: 59
- Left cosets: 60
- Right cosets: 61
- Cosets in abelian groups: 61
- Making things simpler: 61

- Using a different subgroup: 63
- Aside: 64
- Summary: 64

Standard Adjoints 65
- Aside: 65
- The standard adjoint representations of dihedral groups: 65
- The standard adjoint representation of D_4: 67
- Oddness and evenness: 67
- Standard adjoint representations of symmetric groups: 69
- Summary: 69

The Important Stuff 70
- The special theory of relativity: 70
- The complex numbers: 71
- The quaternions: 72
- Gauge theory: 72
- The general theory of relativity: 73

Concluding Remarks 74

Other Books by Dennis Morris 76

Index 80

Introduction

Remarkably, it seems that no book whose subject matter is permutation matrices has ever before been written. There is certainly no such book offered for sale by the internet book-sellers. This small book is written to fill that gap in the literature. This absence from the literature is all the more remarkable given that permutation matrices underlie finite group theory and every type of number; that is every type of division algebra. The whole of trigonometry, the whole of geometry, special relativity, and even the gauge theory of modern theoretical physics is based upon permutation matrices. The real numbers are based on the single 1×1 permutation matrix; the complex numbers, $\mathbb{C}$, are based upon the two 2×2 permutation matrices, and the quaternions are based upon the symmetric 4×4 permutation matrices. The other permutation matrices similarly underlie less well known types of division algebra.

The mathematics of permutation matrices is childishly simple. With a little help, a bright ten-year-old could understand most of this book. Perhaps established mathematicians deem permutation matrices to be too simple to warrant their publishing efforts. Perhaps your author, who has previously written and published sixteen books on higher mathematics and theoretical physics, ought not to be wasting his time writing this simple book on this simple subject. None-the-less, your author feels that, in spite of their simplicity, permutation matrices are of immense importance in mathematics and certainly worthy of a book of their own. Therefore, we will proceed.

The Little Book of Permutation Matrices

Chapter 1

Permutation Matrices

A permutation matrix is a square matrix with a single 1 in each column and a single 1 in each row and zeros everywhere else. For example, a 3×3 permutation matrix is:

$$\begin{bmatrix} 0 & 1 & 0 \\ 0 & 0 & 1 \\ 1 & 0 & 0 \end{bmatrix} \qquad (1.1)$$

There are $n! = n\times(n-1)\times(n-2)\times...\times 3\times 2\times 1$ (n-factorial) different permutation matrices[1] for any given particular size of matrix, $n\times n$; there are $2! = 2$ different 2×2 permutation matrices; there are $3! = 6$ different 3×3 permutation matrices, there are $4! = 24$ different 4×4 permutation matrices, and there are $5! = 120$ different 5×5 permutation matrices etc..

Permutations:
A permutation of n objects – think n differently coloured balls – is just the order, left-to right[2], in which those objects – coloured balls – are arranged. For example, we might have:

$$\begin{array}{ccc} \text{Red} & \text{Green} & \text{Blue} \\ \circ & \circ & \circ \end{array} \qquad (1.2)$$

or, we might have:

---

[1] See below.
[2] Left-to-right is arbitrary. The order could be right-to-left or top-to-bottom or any other order you care to choose.

$$\begin{array}{ccc} \text{Blue} & \text{Green} & \text{Red} \\ \circ & \circ & \circ \end{array} \qquad (1.3)$$

The two different arrangements of the coloured balls, (1.2) & (1.3), are two of the six possible differently ordered arrangements of three coloured balls. We say that the above, (1.2) & (1.3), are two of the six possible permutations of three objects.

Permutation matrices:

Permutation matrices of size $n \times n$ are in a one-to-one correspondence with permutations of the order of n objects. This is just a posh way of saying there is one and only one $n \times n$ permutation matrix corresponding to each and every permutation of the order of n objects.

Mathematicians often write permutations of n objects as two rows of n numbers. The top row is just numbers in the standard order $1, 2, 3, 4, \ldots$ from left to right. The bottom row is the position to which the object – coloured ball – is moved by the permutation. For example, the permutation in which object A is moved from position 1 to position 3 and object B is moved from position 2 to position 2 and object C is moved from position 3 to position 1 is written as:

$$\begin{array}{ccc} A & B & C \\ \downarrow & & \\ C & B & A \end{array} \equiv \begin{array}{ccc} 1 & 2 & 3 \\ \downarrow & & \\ 3 & 2 & 1 \end{array} \qquad (1.4)$$

We read this, (1.4), as one goes to three, two goes to two, and three goes to one. This, (1.4), means: take the object in position 1 as counted from the left and place that object in position 3 as counted from the left and take the object in position 2 as counted from the left and place that object in position 2 as counted from the left and take the object in position 3 as counted from the left and place that object in position 1 as counted from the left. The conventional notation is to omit the $\downarrow$ and to write only the two rows of numbers.

In general, any permutation of n objects can be written uniquely as a $n \times n$ permutation matrix and every permutation matrix can be written uniquely as a permutation; this is exactly what we mean when we say that the permutation matrices are in one-to-one correspondence with the permutations.

We can write the above permutation, (1.4), as a permutation matrix. The permutation matrix will be a 3×3 matrix because there are three objects which are being permuted. We take the top row of numbers in the permutation to be the column counting from left to right of the permutation matrix in which we place the number 1, and we take the bottom row of numbers in the permutation to be the row of the matrix counting from top to bottom in which we place the number 1 in the matrix column corresponding to the top row of numbers in the permutation. We have a couple of examples:

$$\begin{array}{cccc} column & 1\ 2\ 3 & & 1\ 2\ 3 \\ row & 3\ 2\ 1 & & 3\ 1\ 2 \\ & \downarrow & & \downarrow \end{array}$$

$$\begin{bmatrix} 0 & 0 & 1 \\ 0 & 1 & 0 \\ 1 & 0 & 0 \end{bmatrix} \qquad \begin{bmatrix} 0 & 1 & 0 \\ 0 & 0 & 1 \\ 1 & 0 & 0 \end{bmatrix} \qquad (1.5)$$

We see immediately that there is a $n \times n$ permutation matrix for each and every permutation of n objects - the permutation matrices are in a one-to-one correspondence with permutations.

The order of the top row of numbers (or letters) in a permutation does not matter. Only the correlation between the two numbers in a given position in the rows matters. Since it is neater to keep the top row of numbers in ascending order, we usually do this.

Aside:
Instead of taking the top row of the permutation to be the column and the bottom row of the permutation to be the row, we could have taken

the top row of the permutation to be the row and the bottom row of the permutation to be the column:

$$
\begin{array}{cccc}
row & 1\ 2\ 3 & & 1\ 2\ 3 \\
column & 3\ 2\ 1 & & 3\ 1\ 2 \\
& \downarrow & & \downarrow \\
& \begin{bmatrix} 0 & 0 & 1 \\ 0 & 1 & 0 \\ 1 & 0 & 0 \end{bmatrix} & & \begin{bmatrix} 0 & 0 & 1 \\ 1 & 0 & 0 \\ 0 & 1 & 0 \end{bmatrix}
\end{array} \quad (1.6)
$$

Both ways of associating permutation matrices with permutations work equally well, but we must not mix them together. When doing calculations with permutation matrices, it is important to choose only one of the two above procedures, (1.5) & (1.6), to associate permutation matrices with permutations. Different authors choose differently – so be wary.

This alternative procedure, (1.6), produces the inverse[3] permutation matrix[4] of the earlier procedure, (1.5).

The identity permutation:
There is one permutation of n objects which leaves every object in its original position. This 'special' permutation is called the identity permutation. We have the identity permutation of four objects:

$$
\begin{array}{c} 1\ 2\ 3\ 4 \\ 1\ 2\ 3\ 4 \end{array} \equiv \begin{bmatrix} 1 & 0 & 0 & 0 \\ 0 & 1 & 0 & 0 \\ 0 & 0 & 1 & 0 \\ 0 & 0 & 0 & 1 \end{bmatrix} \quad (1.7)
$$

---

[3] The inverse of a permutation matrix, P, is the permutation matrix, P^{-1} such that $PP^{-1} = P^{-1}P = I$ where I is the identity matrix.
[4] All permutation matrices are non-singular and hence have an inverse. The determinant of a permutation matrix is either plus unity or minus unity – see later.

We see that the identity permutation corresponds to the identity matrix. This is true what-ever way we choose to associate a permutation matrix with a permutation.

Sequential combination of permutations:

We can begin with a set of objects in a given order and change that order. This is a permutation of the order of the objects, of course. We can then start again with the objects in the new order and once again re-order the objects:

$$
\begin{array}{ccc} A & B & C \\ & \downarrow & \\ B & C & A \\ & \downarrow & \\ C & A & B \end{array} \qquad \begin{array}{ccc} 1 & 2 & 3 \\ 2 & 3 & 1 \\ & \& & \\ 2 & 3 & 1 \\ 3 & 1 & 2 \end{array} \qquad (1.8)
$$

This combination of two permutations, (1.8), is the same as a single permutation:

$$
\begin{array}{ccc} A & B & C \\ & \downarrow & \\ C & A & B \end{array} \qquad \begin{array}{ccc} 1 & 2 & 3 \\ 3 & 1 & 2 \end{array} \qquad (1.9)
$$

Of course, what-so-ever permutations we choose to sequentially combine, we will get another permutation of the objects. We say that full set of permutations are closed under sequential combination.

A note:

We would normally write the bottom rightmost of the above permutations, (1.8), as:

$$
\begin{array}{ccc} 2 & 3 & 1 \\ 3 & 1 & 2 \end{array} = \begin{array}{ccc} 1 & 2 & 3 \\ 2 & 3 & 1 \end{array} \qquad (1.10)
$$

These, two permutations, (1.10), are the same permutation written differently; this notation does no more than signify which positions change.

The conventional notation - combining permutations:
The sequential combination of permutations is usually written horizontally rather than vertically, and the sequential combination is customarily, but arbitrarily, done from right to left. Working from the right towards the left, we follow each object through both permutations:

$$\begin{pmatrix} 1 & 2 & 3 \\ 3 & 1 & 2 \end{pmatrix} \stackrel{\leftarrow}{=} \begin{pmatrix} 1 & 2 & 3 \\ 2 & 3 & 1 \end{pmatrix}\begin{pmatrix} 1 & 2 & 3 \\ 2 & 3 & 1 \end{pmatrix}$$

$$3 \leftarrow 1 \stackrel{\leftarrow}{=} 3 \leftarrow 2 \leftarrow 1 \qquad (1.11)$$
$$1 \leftarrow 2 \stackrel{\leftarrow}{=} 1 \leftarrow 3 \leftarrow 2$$
$$2 \leftarrow 3 \stackrel{\leftarrow}{=} 2 \leftarrow 1 \leftarrow 3$$

We see that the permutation on the left of (1.11) is the same permutation as we calculated in (1.8) when we combined the two permutations. The above notation and right-to-left sequential combination of permutations is standard in mathematics. It might surprise the reader to know that such sequential combination of permutations, (1.11), is 'really' matrix multiplication done from left-to-right. We have the above two permutations, (1.11), written as permutation matrices; we multiply them together using standard left-to-right matrix multiplication[5]:

$$\begin{bmatrix} 0 & 0 & 1 \\ 1 & 0 & 0 \\ 0 & 1 & 0 \end{bmatrix}\begin{bmatrix} 0 & 0 & 1 \\ 1 & 0 & 0 \\ 0 & 1 & 0 \end{bmatrix} \stackrel{\rightarrow}{=} \begin{bmatrix} 0 & 1 & 0 \\ 0 & 0 & 1 \\ 1 & 0 & 0 \end{bmatrix} \qquad (1.12)$$

Looking at (1.11), we see that matrix multiplication from left-to-right, (1.12), is 'really' the sequential combination of permutations from right-to-left. Well, it is for permutation matrices. Of course, the left-to-

---

[5] We multiply the elements in each row of the left-most matrix by the elements in each column in turn of the middle matrix and add these products to form the element in the product matrix on the right. For permutation matrices, most of the products are zero.

right nature of matrix multiplication is no more than a well-established and agreed arbitrary custom.

Aside:
We have just presented the standard convention for combining permutations from right-to-left, (1.11). We have multiplied together the two permutation matrices, (1.12), in the standard left-to-right convention. Now, suppose we had formed the permutation matrices taking the top row of the permutation to be the row of the matrix rather than the column of the matrix as we presented in the aside above, circa (1.6); we would have:

$$\begin{pmatrix} 1 & 2 & 3 \\ 2 & 3 & 1 \end{pmatrix} \quad \begin{pmatrix} 1 & 2 & 3 \\ 3 & 2 & 1 \end{pmatrix}$$
$$\downarrow \quad\quad\quad \downarrow \quad\quad\quad (1.13)$$
$$\begin{bmatrix} 0 & 1 & 0 \\ 0 & 0 & 1 \\ 1 & 0 & 0 \end{bmatrix} \quad \begin{bmatrix} 0 & 0 & 1 \\ 0 & 1 & 0 \\ 1 & 0 & 0 \end{bmatrix}$$

If we combine the two permutations together from left-to-right rather than the conventional way of right-to-left, we get:

$$\begin{pmatrix} 1 & 2 & 3 \\ 2 & 3 & 1 \end{pmatrix} \rightarrow \begin{pmatrix} 1 & 2 & 3 \\ 3 & 2 & 1 \end{pmatrix} = \begin{pmatrix} 1 & 2 & 3 \\ 2 & 1 & 3 \end{pmatrix} \quad (1.14)$$

This, (1.14), corresponds, under this non-conventional notation, to multiplying the two corresponding matrices together left-to-right:

$$\begin{bmatrix} 0 & 1 & 0 \\ 0 & 0 & 1 \\ 1 & 0 & 0 \end{bmatrix} \begin{bmatrix} 0 & 0 & 1 \\ 0 & 1 & 0 \\ 1 & 0 & 0 \end{bmatrix} = \begin{bmatrix} 0 & 1 & 0 \\ 1 & 0 & 0 \\ 0 & 0 & 1 \end{bmatrix} \quad (1.15)$$

The reader might think it would be better to use this notation rather than the conventional notation. The reader might be correct, but we are stuck with the conventions. I can assure the reader that nature cares not a jot for how we denote our sequential combination of permutations and how we correspond our permutation matrices with our permutations.

Multiplicative closure:
It is a mathematical fact that the product of two permutation matrices (by conventional matrix multiplication) is another permutation matrix. This follows from the fact that two permutations sequentially combined together must produce another permutation.

Of course, if we do matrix multiplication a step at a time, with thought, we see that each step of multiplying together two permutation matrices will produce a single 1. With more thought, we see that these 1's will never be on the same row or in the same column as each other.

Matrix multiplication:
Within mathematics, there is 'genuine' multiplication and there are calculative procedures which are sloppily called multiplication. 'Genuine' multiplication exists only within a division algebra[6] and has properties like multiplicative closure[7], absence of zero divisors etc.. The multiplication together of two permutation matrices is 'genuine' multiplication. Not all 'matrix multiplication' is 'genuine' multiplication; an example of 'not genuine' multiplication is:

$$\begin{bmatrix} a & b & c \\ d & e & f \end{bmatrix} \begin{bmatrix} r & s \\ t & u \\ v & w \end{bmatrix} = \begin{bmatrix} ar+bt+cv & as+bu+cw \\ dr+et+fv & ds+eu+fw \end{bmatrix} \quad (1.16)$$

This, (1.16), is not 'genuine' multiplication; it is a duck mated with a chicken, and it has produced a robin.

---

[6] Division algebras are types of numbers like the real numbers, the complex numbers, the quaternions, or any of an infinite number of such types of numbers.
[7] Multiplicative closure says that you can multiply together only objects of the same nature and the product must be of that same nature – a duck mated with a duck produces a duck; you cannot mate a duck with a frog, and there would be something wrong with a pair of ducks if, when mated, they produced a mouse.

At school, the reader will have been taught matrix multiplication without any explanation of why matrices are multiplied together in the way we are taught. Perhaps someone might have mentioned linear multiplication. 'Genuine' matrix multiplication, that is multiplication within division algebras, is no more than a 'beefed up' version of the sequential combination of permutations – that's quite profound.

We mentioned above that 'genuine' multiplication exists in only a division algebra and that the multiplication of permutation matrices is such 'genuine' multiplication. The alert reader will realise that this implies permutation matrices must be elements of a division algebra. This is true; a permutation matrix corresponds to a variable within the division algebra which is equal to 1^8. The permutation matrices are the skeleton upon which division algebras are built – more later.

Inverse permutation matrices:

For every permutation, there is a 'reverse permutation' which will undo the original permutation; for example:

$$\begin{matrix} A & B & C \\ C & A & B \end{matrix} \equiv \begin{matrix} 1 & 2 & 3 \\ 3 & 1 & 2 \end{matrix}$$

&

$$\begin{matrix} C & A & B \\ A & B & C \end{matrix} \equiv \begin{matrix} A & B & C \\ B & C & A \end{matrix} \equiv \begin{matrix} 1 & 2 & 3 \\ 2 & 3 & 1 \end{matrix} \quad (1.17)$$

$$\begin{pmatrix} 1 & 2 & 3 \\ 1 & 2 & 3 \end{pmatrix} = \begin{pmatrix} 1 & 2 & 3 \\ 3 & 1 & 2 \end{pmatrix} \begin{pmatrix} 1 & 2 & 3 \\ 2 & 3 & 1 \end{pmatrix}$$

When a permutation and its 'reverse permutation' are sequentially combined, in any order, the permutation will be undone and we get back to the identity permutation. The formal name of a 'reverse permutation'

---

[8] The identity permutation matrix corresponds to the real variable. Non-identity permutation matrices each correspond to an imaginary variable.

is the inverse permutation. It is easy to find the inverse of a given permutation; we simply swap the two rows:

$$\text{The inverse of } \begin{pmatrix} 1 & 2 & 3 \\ 3 & 1 & 2 \end{pmatrix} \text{ is } \begin{pmatrix} 3 & 1 & 2 \\ 1 & 2 & 3 \end{pmatrix} = \begin{pmatrix} 1 & 2 & 3 \\ 2 & 3 & 1 \end{pmatrix} \quad (1.18)$$

Written as permutation matrices, the above two permutations, (1.17), are:

$$\begin{pmatrix} 1 & 2 & 3 \\ 1 & 2 & 3 \end{pmatrix} = \begin{pmatrix} 1 & 2 & 3 \\ 3 & 1 & 2 \end{pmatrix} \begin{pmatrix} 1 & 2 & 3 \\ 2 & 3 & 1 \end{pmatrix}$$

$$\begin{bmatrix} 0 & 1 & 0 \\ 0 & 0 & 1 \\ 1 & 0 & 0 \end{bmatrix} \begin{bmatrix} 0 & 0 & 1 \\ 1 & 0 & 0 \\ 0 & 1 & 0 \end{bmatrix} = \begin{bmatrix} 1 & 0 & 0 \\ 0 & 1 & 0 \\ 0 & 0 & 1 \end{bmatrix} \quad (1.19)$$

The product of a permutation matrix and its inverse is the identity matrix; indeed, this is the definition of an inverse permutation matrix.

Since the top row of a permutation gives the matrix column in which to place the 1 and the bottom row of a permutation gives the matrix row in which to place the 1, when we calculate the inverse of a permutation by swapping the two rows of the permutation, we are swapping rows for columns in the permutation matrix. This operation of swapping the rows and columns of a permutation matrix is called transposition. Matrix transposition is most easily visualised as reflection across the leading diagonal[9] of the matrix. We write the matrix inverse with a superscript (-1). We have:

$$\begin{bmatrix} 0 & 1 & 0 \\ 0 & 0 & 1 \\ 1 & 0 & 0 \end{bmatrix}^{-1} = \begin{bmatrix} 0 & 1 & 0 \\ 0 & 0 & 1 \\ 1 & 0 & 0 \end{bmatrix}^{Transpose} = \begin{bmatrix} 0 & 0 & 1 \\ 1 & 0 & 0 \\ 0 & 1 & 0 \end{bmatrix} \quad (1.20)$$

---

[9] The leading diagonal of a matrix runs from the top left-hand corner of the matrix to the bottom right-hand corner of the matrix.

Simply, the inverse of a permutation matrix is the transpose of that permutation matrix. This is not the case for matrices in general; remember, only permutation matrices.

$$P^{-1} = P^{Transpose} \qquad\qquad PP^{-1} = P^{-1}P = I \qquad (1.21)$$

The existence of a non-zero inverse, transpose, for every permutation matrix implies the permutation matrices are all non-singular.

The non-commutativity of permutation matrices:
It is an easily observable fact that not all pairs of permutation matrices commute. By this, we mean that the order of multiplication is important. We have:

$$\begin{bmatrix} 1 & 0 & 0 \\ 0 & 0 & 1 \\ 0 & 1 & 0 \end{bmatrix} \begin{bmatrix} 0 & 0 & 1 \\ 0 & 1 & 0 \\ 1 & 0 & 0 \end{bmatrix} = \begin{bmatrix} 0 & 0 & 1 \\ 1 & 0 & 0 \\ 0 & 1 & 0 \end{bmatrix} \qquad (1.22)$$

But, swapping the order of the two matrices:

$$\begin{bmatrix} 0 & 0 & 1 \\ 0 & 1 & 0 \\ 1 & 0 & 0 \end{bmatrix} \begin{bmatrix} 1 & 0 & 0 \\ 0 & 0 & 1 \\ 0 & 1 & 0 \end{bmatrix} = \begin{bmatrix} 0 & 1 & 0 \\ 0 & 0 & 1 \\ 1 & 0 & 0 \end{bmatrix} \qquad (1.23)$$

Some pairs of permutation matrices do commute – the identity commutes with every permutation matrix. This non-commutativity of permutation matrices is exactly the non-commutativity of permutations in general.

How many permutation matrices?:
How many $n \times n$ permutation matrices are there for a given n? Well, within a $n \times n$ matrix, there are n positions on the top row in which we can place a 1. There are $(n-1)$ positions on the second row in which we can place a 1; remember, we cannot have two 1's in the same column. Similarly, there are $(n-2)$ positions on the third row in which

we can place a 1. There is only one position on the bottom row in which we can place a 1. Thus there are:

$$n(n-1)(n-2)...(2)(1) = n! \qquad (1.24)$$

permutation matrices of size $n \times n$. To reiterate, there are $n!$ different permutation matrices of size $n \times n$.

Summary:

The permutation matrices of a given size are a multiplicatively closed set.

The inverse of a permutation matrix is the transpose of that permutation matrix; this implies that all permutation matrices, P, are non-singular.

In every set of $n \times n$ permutation matrices, there is an identity permutation matrix which is the matrix identity.

Permutation matrices are not necessarily commutative.

Addendum:

Permutation matrices will permute the rows or columns of other matrices; for example:

$$\begin{bmatrix} 0 & 1 & 0 \\ 0 & 0 & 1 \\ 1 & 0 & 0 \end{bmatrix} \begin{bmatrix} a & b & c \\ d & e & f \\ g & h & i \end{bmatrix} = \begin{bmatrix} d & e & f \\ g & h & i \\ a & b & c \end{bmatrix}$$

&

$$\begin{bmatrix} a & b & c \\ d & e & f \\ g & h & i \end{bmatrix} \begin{bmatrix} 0 & 1 & 0 \\ 0 & 0 & 1 \\ 1 & 0 & 0 \end{bmatrix} = \begin{bmatrix} c & a & b \\ f & d & e \\ i & g & h \end{bmatrix} \qquad (1.25)$$

It might be that permutation matrices were so named because of this property.

The Little Book of Permutation Matrices

Chapter 2

Determinants and Permutation Matrices

This chapter is a long smouldering chapter. Throughout this chapter, we develop many seemingly unrelated concepts; we then bring these concepts together at the end of the chapter.

We begin this long smouldering chapter with a digression presenting some notation.

Digression - the Levi-Civita symbol – notation only:
The Levi-Civta symbol is:

$$\varepsilon_{abcd\ldots} \tag{2.1}$$

The subscripts are ordered; this means that the order of the subscripts is important. We have: $\varepsilon_{abcd\ldots} \neq \varepsilon_{bacd\ldots}$ wherein we have swapped the order of two of the subscripts.

The number of subscripts depends upon the context in which the Levi-Civita symbol is being used; this is the size of the permutation matrix for our present purposes.

When the subscripts are in the proper order, the Levi-Civita symbol is equal to plus unity.

$$\varepsilon_{abc} = +1 \tag{2.2}$$

Every time a single subscript swaps order with its immediate neighbour, the sign of the Levi-Civita symbol changes. We have:

$$\begin{array}{llll} \varepsilon_{abc} = +1, & \varepsilon_{acb} = -1, & \varepsilon_{bac} = -1, & \varepsilon_{bca} = +1, \\ & \varepsilon_{cab} = +1, & \varepsilon_{cba} = -1 & \end{array} \tag{2.3}$$

If a subscript is repeated, the Levi-Civita is zero:

Determinants and Permutation Matrices

$$\varepsilon_{aac} = 0, \quad \varepsilon_{bcb} = 0 \tag{2.4}$$

The Levi-Civita symbol is not a mathematical entity; it is merely useful and commonly used notation.

End of digression - now to determinants:
Every square matrix has associated with it a number[10] called the determinant of that matrix. The determinant is a sum of particular products of the elements of the matrix; we will be using only real numbers as elements of our matrices, and so the determinant of our matrices will always be a real number.

Only square matrices have a determinant.

If the determinant of a matrix is zero, then that matrix is said to be singular. Singular matrices have no inverse. Other than this mention, singular matrices are of no interest to us.

The determinant of a matrix is 'an invariant' of that matrix. This means that, if we were to write the matrix in a different basis (by using a similarity transformation), then the determinant would be unchanged by the similarity transformation – the determinant is invariant under change of basis. Of course, choice of basis is the arbitrary choice of humankind. The determinant of a matrix is above the arbitrariness of human whim.

The matrix determinant:
The determinant of a $n \times n$ matrix is formed from all possible products of n elements of the matrix such that each of the n elements in each product are from a different row of the matrix and are from a different column of the matrix. For example, there are only two ways to combine two of the four elements of a 2×2 matrix such that each of the pair of

---

[10] A number is an element of a division algebra; for example, the determinant might be a complex number if the elements of the matrix are complex numbers.

elements are in a different row of the matrix and are in a different column of the matrix; we have:

$$\begin{bmatrix} a & b \\ c & d \end{bmatrix} \qquad (2.5)$$

$$\{ad, bc\}$$

Similarly there are six such combinations for a 3×3 matrix:

$$\begin{bmatrix} a & b & c \\ d & e & f \\ g & h & i \end{bmatrix} \qquad (2.6)$$

$$\{aei, bfg, cdh, ceg, bdi, afh\}$$

In a $n \times n$ matrix, there are $n!$ such products of n elements. The determinant of the matrix is formed from these $n!$ products of n elements by addition and subtraction. For example, the determinant of the 2×2 matrix, (2.5), is:

$$\det\left(\begin{bmatrix} a & b \\ c & d \end{bmatrix}\right) = ab - cd \qquad (2.7)$$

The determinant of the 3×3 matrix, (2.6), is:

$$\det\left(\begin{bmatrix} a & b & c \\ d & e & f \\ g & h & i \end{bmatrix}\right) = aei - afh + bfg - bdi + cdh - ceg \qquad (2.8)$$

We have to watch the minus signs.

Let us impose the 'shadows' of the six 3×3 permutation matrices upon the matrix (2.6). We have:

Determinants and Permutation Matrices

$$\begin{bmatrix} 1 & 0 & 0 \\ 0 & 1 & 0 \\ 0 & 0 & 1 \end{bmatrix} \begin{bmatrix} 0 & 1 & 0 \\ 0 & 0 & 1 \\ 1 & 0 & 0 \end{bmatrix} \begin{bmatrix} 0 & 0 & 1 \\ 1 & 0 & 0 \\ 0 & 1 & 0 \end{bmatrix}$$

$$\begin{bmatrix} a & \sim & \sim \\ \sim & e & \sim \\ \sim & \sim & i \end{bmatrix} \begin{bmatrix} \sim & b & \sim \\ \sim & \sim & f \\ g & \sim & \sim \end{bmatrix} \begin{bmatrix} \sim & \sim & c \\ d & \sim & \sim \\ \sim & h & \sim \end{bmatrix}$$

(2.9)

$$\begin{bmatrix} 1 & 0 & 0 \\ 0 & 0 & 1 \\ 0 & 1 & 0 \end{bmatrix} \begin{bmatrix} 0 & 0 & 1 \\ 0 & 1 & 0 \\ 1 & 0 & 0 \end{bmatrix} \begin{bmatrix} 0 & 1 & 0 \\ 1 & 0 & 0 \\ 0 & 0 & 1 \end{bmatrix}$$

$$\begin{bmatrix} a & \sim & \sim \\ \sim & \sim & f \\ \sim & h & \sim \end{bmatrix} \begin{bmatrix} \sim & \sim & c \\ \sim & e & \sim \\ g & \sim & \sim \end{bmatrix} \begin{bmatrix} \sim & b & \sim \\ d & \sim & \sim \\ \sim & \sim & i \end{bmatrix}$$

(2.10)

Looking at (2.6) or (2.8), we see that the permutation matrices in (2.9) & (2.10) each 'pick out' the elements which are to be multiplied together to form the determinant; this, using the appropriately sized permutation matrices, is general for all square matrices of any size.

We now have each term of the determinant, but still have to find the signs which precede each term.

Back to the Levi-Civita symbol:
The standard way of expressing the signs in a determinant of a matrix is to use the Levi-Civita symbol to indicate the sign, plus or minus, which precedes each term; for example:

$$\det\left(\begin{bmatrix} a & b & c \\ d & e & f \\ g & h & i \end{bmatrix} \right) = \quad (2.11)$$

$$\varepsilon_{123}aei + \varepsilon_{132}afh + \varepsilon_{231}bfg + \varepsilon_{213}bdi + \varepsilon_{312}cdh + \varepsilon_{321}ceg$$

The subscripts of the Levi-Civita symbol are the columns of the elements in the particular product in order starting with the top row and working downward; for example, the term:

$$bfg \to \varepsilon_{231} \qquad (2.12)$$

The subscripts correspond to the column of b in the top row, which is column 2 from the left, followed by the column of f in the second row down, which is column 3 from the left, followed by the column of g in the bottom row, which is 1 from the left. Yes, it's a little confusing and very error prone. There is a better way.

The Levi-Civita symbol and permutation matrices:
We can view any permutation matrix as a copy of the identity matrix with a pair, or more than one pair, of swapped columns:

$$\begin{bmatrix} 1 & 0 & 0 \\ 0 & 1 & 0 \\ 0 & 0 & 1 \end{bmatrix} \xrightarrow{\text{Swap col 2 for col 3}} \begin{bmatrix} 1 & 0 & 0 \\ 0 & 0 & 1 \\ 0 & 1 & 0 \end{bmatrix} \qquad (2.13)$$

We can associate this swapping of columns with the swapping of subscripts of the Levi-Civita symbol:

$$\begin{bmatrix} 1 & 0 & 0 \\ 0 & 1 & 0 \\ 0 & 0 & 1 \end{bmatrix} \xrightarrow{\text{Swap col 2 for col 3}} \begin{bmatrix} 1 & 0 & 0 \\ 0 & 0 & 1 \\ 0 & 1 & 0 \end{bmatrix}$$

$$\varepsilon_{123} \qquad \text{Swap subscript 2 for subscript 3} \qquad \varepsilon_{132}$$

$$(2.14)$$

We have attached a Levi-Civita symbol to each matrix in (2.14) wherein the subscripts are the order of the columns based upon the initial order being the identity matrix. We can do this with every permutation matrix. We can do this with permutation matrices of any size. There are six 3×3 permutation matrices; these can all be formed from the identity matrix by swapping columns. We have:

Determinants and Permutation Matrices

$$\begin{bmatrix} 1 & 0 & 0 \\ 0 & 1 & 0 \\ 0 & 0 & 1 \end{bmatrix} \begin{bmatrix} 0 & 1 & 0 \\ 0 & 0 & 1 \\ 1 & 0 & 0 \end{bmatrix} \begin{bmatrix} 0 & 0 & 1 \\ 1 & 0 & 0 \\ 0 & 1 & 0 \end{bmatrix} \quad (2.15)$$
$$\varepsilon_{123} \qquad\qquad \varepsilon_{231} \qquad\qquad \varepsilon_{312}$$

And:

$$\begin{bmatrix} 1 & 0 & 0 \\ 0 & 0 & 1 \\ 0 & 1 & 0 \end{bmatrix} \begin{bmatrix} 0 & 0 & 1 \\ 0 & 1 & 0 \\ 1 & 0 & 0 \end{bmatrix} \begin{bmatrix} 0 & 1 & 0 \\ 1 & 0 & 0 \\ 0 & 0 & 1 \end{bmatrix} \quad (2.16)$$
$$\varepsilon_{132} \qquad\qquad \varepsilon_{321} \qquad\qquad \varepsilon_{213}$$

We have a correspondence between the permutation matrices and the Levi-Civita symbol. This correspondence is between the swapped columns of the permutation matrices and the swapped subscripts of the Levi-Civita symbol. We can thus use the permutation matrices instead of the Levi-Civita symbol - we simply count the column swaps to get the sign. It works, but let us continue.

The determinants of permutation matrices:
Since they are square matrices, permutation matrices have a determinant. With our eyes on the above, (2.11), and a little thought, we realise that the determinant of a permutation matrix will always be of the form:

$$\det\left(\begin{bmatrix} 0 & 1 & 0 & 0 \\ 0 & 0 & 0 & 1 \\ 0 & 0 & 1 & 0 \\ 1 & 0 & 0 & 0 \end{bmatrix}\right) = \varepsilon_{a..d} 1.1.1.1 = \varepsilon_{a..d} = \pm 1 \quad (2.17)$$

We see that permutation matrices always have a determinant of plus unity or minus unity.

It is an established fact that swapping two columns of a square matrix will change the sign of the determinant. Looking at the above, we have, in a long-winded way, shown this to be the case. Thus, we can calculate the sign of the determinant of a permutation matrix by counting the number of column swaps needed to return the permutation matrix to the identity.

We begin with the identity matrix, whose determinant is +1, and swap columns to form a different permutation matrix. When the number of column swaps is odd, the determinant of the different permutation matrix is −1. When the number of column swaps is even, the determinant of the different permutation matrix is +1. We see, (2.17), that we can replace the Levi-Civita symbol in (2.11) by the determinant of a permutation matrix.

The matrix determinant written using permutation matrices:
We have now reached the end of our long smouldering; we have:

$$\det\left(\begin{bmatrix} a & b & c \\ d & e & f \\ g & h & i \end{bmatrix}\right) =$$

$$\det\left(\begin{bmatrix} 1 & 0 & 0 \\ 0 & 1 & 0 \\ 0 & 0 & 1 \end{bmatrix}\right) \Pr\left(\begin{bmatrix} a & 0 & 0 \\ 0 & e & 0 \\ 0 & 0 & i \end{bmatrix}\right) + \det\left(\begin{bmatrix} 1 & 0 & 0 \\ 0 & 0 & 1 \\ 0 & 1 & 0 \end{bmatrix}\right) \Pr\left(\begin{bmatrix} a & 0 & 0 \\ 0 & 0 & f \\ 0 & h & 0 \end{bmatrix}\right) +$$

$$+ \det\left(\begin{bmatrix} 0 & 0 & 1 \\ 0 & 1 & 0 \\ 1 & 0 & 0 \end{bmatrix}\right) \Pr\left(\begin{bmatrix} 0 & 0 & c \\ 0 & e & 0 \\ f & 0 & 0 \end{bmatrix}\right) + \det\left(\begin{bmatrix} 0 & 1 & 0 \\ 1 & 0 & 0 \\ 0 & 0 & 1 \end{bmatrix}\right) \Pr\left(\begin{bmatrix} 0 & b & 0 \\ d & 0 & 0 \\ 0 & 0 & i \end{bmatrix}\right)$$

$$+ \det\left(\begin{bmatrix} 0 & 1 & 0 \\ 0 & 0 & 1 \\ 1 & 0 & 0 \end{bmatrix}\right) \Pr\left(\begin{bmatrix} 0 & b & 0 \\ 0 & 0 & f \\ g & 0 & 0 \end{bmatrix}\right) + \det\left(\begin{bmatrix} 0 & 0 & 1 \\ 1 & 0 & 0 \\ 0 & 1 & 0 \end{bmatrix}\right) \Pr\left(\begin{bmatrix} 0 & 0 & c \\ d & 0 & 0 \\ 0 & h & 0 \end{bmatrix}\right)$$

$$= +1.aei - 1.afh - 1.cef - 1.bdi + 1.bfg + 1.cdh$$

(2.18)

Determinants and Permutation Matrices

In this, (2.18), the Pr([ ]) means take the product of the elements in the matrix. Counting the column swaps will give the determinant of the permutation matrices which is the sign preceding each product.

We see that the determinant uses the whole set of permutation matrices.

In general, the determinant of any square matrix is calculated, using the appropriately sized permutation matrices, in the same way as we have above, (2.18). So, if you cannot remember the formula for the determinant of a $n \times n$ matrix and cannot with confidence untangle the general formula in the $n \times n$ case, you can always use the above (2.18) formula to calculate a determinant.

Note that we could have done the above by swapping rows rather than by swapping columns. The result would have been the same.

General properties of determinants:
For the sake of completeness, we list a few properties of determinants. We have:

a) The determinant of a matrix is unique[11].
b) $\det(A^{Transpose}) = \det(A)$
c) $\det(A)\det(B) = \det(AB)$
d) $\det(cA) = c^n \det(A) : c \in \mathbb{R}$ and n is the size of the matrix
e) The determinant of a matrix is equal to the product of the eigenvalues of the matrix.

The trace of a matrix is the sum of the elements on the leading diagonal. The trace of a matrix is the sum of the eigenvalues of the matrix.

The nature of the determinant:
The determinant of a $n \times n$ matrix is often interpreted as the n-dimensional 'volume' of a n-dimensional parallelepiped constructed by taking the rows of the matrix to be vectors. This interpretation assumes

---

[11] Serge Lang: Linear Algebra 2nd edition 1971 pp 173 & 191.

a Euclidean type of space with a Riemann distance function, and so it is questionable.

Within division algebras[12], the determinant is the distance function of the space; an example is the 2-dimensional complex numbers, $\mathbb{C}$:

$$\det\left(\begin{bmatrix} a & b \\ -b & a \end{bmatrix}\right) = \det\left(\begin{bmatrix} r & 0 \\ 0 & r \end{bmatrix}\begin{bmatrix} \cos\theta & \sin\theta \\ -\sin\theta & \cos\theta \end{bmatrix}\right) \quad (2.19)$$

$$a^2 + b^2 = r^2$$

This 'distance function' property of determinants is universal and not dependent upon the nature of the space.

Aside:
We often see matrices with complex elements, $\mathbb{C}$. We might thus be led to write matrices with quaternion elements; for example:

$$\begin{bmatrix} \mathbb{H}_1 & \mathbb{H}_2 \\ \mathbb{H}_3 & \mathbb{H}_4 \end{bmatrix} \quad (2.20)$$

The determinant of this matrix, (2.20), would be:

$$\det{}_1 = \mathbb{H}_1\mathbb{H}_4 - \mathbb{H}_2\mathbb{H}_3 \quad (2.21)$$

We might equally well define the determinant to be:

$$\det{}_2 = \mathbb{H}_4\mathbb{H}_1 - \mathbb{H}_2\mathbb{H}_3 \quad (2.22)$$

There is a problem. Because quaternions are non-commutative, the two determinants, (2.21) & (2.22), are not equal.

However, a quaternion can be written unambiguously as a 4×4 matrix with real numbers as elements. We present a left-chiral quaternion.[13]

---

[12] All division algebras can be written as matrices. Thus, all division algebras have a polar form which includes a rotation matrix with determinant unity.
[13] There are also right-chiral quaternions – see Dennis Morris : Quaternions.

Determinants and Permutation Matrices

$$\mathbb{H}_{L\chi} = \begin{bmatrix} a & b & c & d \\ -b & a & -d & c \\ -c & d & a & -b \\ -d & -c & b & a \end{bmatrix} \quad (2.23)$$

Using the block multiplication properties of matrices, we can write the matrix with quaternion elements, (2.20), as the 8×8 matrix:

$$\begin{bmatrix} [4\times 4]_1 & [4\times 4]_2 \\ [4\times 4]_3 & [4\times 4]_4 \end{bmatrix} \quad (2.24)$$

The determinant of this 8×8 matrix is uniquely determined. We have a problem. The problem arises from using quaternions as elements of a matrix – we should never do this. Although it is not erroneous to use complex numbers as elements of a matrix, it can lead us astray, and it is not really cricket.

Chapter 3

Odd and Even Permutation Matrices

In the previous chapter, we saw that we can separate all permutation matrices, P, into two types depending upon their determinant – this is two types of permutations. The permutation matrices with $\det(P) = +1$ are called even permutation matrices, and the permutation matrices with $\det(P) = -1$ are called odd permutation matrices. Similar nomenclature is applied to permutations in general. The even permutation matrices are the permutation matrices which differ from the identity matrix by an even number of column swaps. The odd permutation matrices are the permutation matrices which differ from the identity matrix by an odd number of column swaps.

Even permutation matrices are a closed set:
The reader will be aware of the identity:

$$\det(A)\det(B) = \det(AB) \qquad (3.1)$$

Even permutation matrices have determinant $+1$, and odd permutation matrices have determinant -1. We see that:

a) the product of two even permutation matrices is an even permutation matrix
b) the product of two odd permutation matrices is an even permutation matrix
c) the product of an even permutation matrix and an odd permutation matrix is an odd permutation matrix

The inverse of an even permutation is an even permutation:
The product of a matrix and its inverse is the identity matrix. The identity matrix has determinant $+1$. Using (3.1), we see that the inverse

of an even permutation matrix must also be an even permutation matrix because $+1 \times -1 \neq +1$. Similarly, the inverse of an odd permutation matrix must be an odd permutation matrix.

We have seen that the inverse of a permutation matrix is the transpose of that permutation matrix. It is well established mathematics that:

$$\det(A^{Transpose}) = \det(A) \qquad (3.2)$$

Odd for even:
Since odd permutation matrices are derived from the identity matrix by an odd number of column swaps and even permutation matrices are derived from the identity matrix by an even number of column swaps, we can change an even permutation matrix into an odd permutation matrix by a single swapping of two columns and vice-versa.

For any size, $n \times n$, of matrices, there are as many even permutation matrices as there are odd permutation matrices. The identity is one of the even permutation matrices, of course.

Since there are $n!$ $n \times n$ permutation matrices of any given size, n, there are $\dfrac{n!}{2}$ even permutation matrices of any given size and $\dfrac{n!}{2}$ odd permutation matrices of any given size.

Chapter 4

The Order of a Permutation Matrix

Any permutation sequentially combined with itself enough times will eventually produce the identity permutation. The same is true of permutation matrices. Any permutation matrix raised to the appropriate power will be the identity permutation matrix; for example:

$$\begin{bmatrix} 0 & 1 \\ 1 & 0 \end{bmatrix}^2 = \begin{bmatrix} 1 & 0 \\ 0 & 1 \end{bmatrix} \tag{4.1}$$

$$\begin{bmatrix} 0 & 1 & 0 \\ 0 & 0 & 1 \\ 1 & 0 & 0 \end{bmatrix}^3 = \begin{bmatrix} 1 & 0 & 0 \\ 0 & 1 & 0 \\ 0 & 0 & 1 \end{bmatrix} \tag{4.2}$$

We are therefore justified in thinking of $n \times n$ permutation matrices as being various roots of unity. We can think of the matrix in (4.2) as a cube root of plus unity. Although this view of permutations is not currently the fashion, it is the way that the mathematician Arthur Cayley (1821-1895) thought of permutations. There are connections to the higher dimensional types of numbers, like the complex numbers or the quaternions, which we will make clearer in the last chapter of this book.

Aside:
Interestingly, we have:

$$\begin{bmatrix} 0 & 1 \\ -1 & 0 \end{bmatrix}^2 = \begin{bmatrix} -1 & 0 \\ 0 & -1 \end{bmatrix} \tag{4.3}$$

$$\begin{bmatrix} 0 & 1 & 0 \\ 0 & 0 & -1 \\ 1 & 0 & 0 \end{bmatrix}^3 = \begin{bmatrix} -1 & 0 & 0 \\ 0 & -1 & 0 \\ 0 & 0 & -1 \end{bmatrix} \qquad (4.4)$$

We see that a minor modification of the permutation matrix in (4.1) will produce a square root of minus unity and a minor modification of the permutation matrix in (4.2) will produce a cube root of minus unity.

The order of a permutation matrix:

The power to which the permutation matrix must be raised to make it equal to the identity matrix is called the order of the permutation matrix. A permutation matrix of order, say, 47, is a 47th root of plus unity.

A permutation matrix raised to power zero is the identity matrix.

Permutation matrices can be raised to negative powers. Of course, the inverse of a permutation matrix is that permutation matrix raised to the power of -1.

A few examples:

We have the order two permutation matrix:

$$\begin{bmatrix} 1 & 0 & 0 \\ 0 & 0 & 1 \\ 0 & 1 & 0 \end{bmatrix}^2 = \begin{bmatrix} 1 & 0 & 0 \\ 0 & 1 & 0 \\ 0 & 0 & 1 \end{bmatrix} \qquad (4.5)$$

We have the order three permutation matrix:

$$\begin{bmatrix} 0 & 0 & 1 \\ 1 & 0 & 0 \\ 0 & 1 & 0 \end{bmatrix}^3 = \begin{bmatrix} 1 & 0 & 0 \\ 0 & 1 & 0 \\ 0 & 0 & 1 \end{bmatrix} \qquad (4.6)$$

The counter identity matrix[14] of any size is always an order two permutation matrix:

$$\begin{bmatrix} 0 & 0 & 1 \\ 0 & 1 & 0 \\ 1 & 0 & 0 \end{bmatrix}^2 = \begin{bmatrix} 1 & 0 & 0 \\ 0 & 1 & 0 \\ 0 & 0 & 1 \end{bmatrix} \qquad (4.7)$$

The same is true of any symmetric permutation matrix. The inverse of a symmetric permutation matrix is itself:

$$\begin{bmatrix} 0 & 0 & 1 & 0 \\ 0 & 0 & 0 & 1 \\ 1 & 0 & 0 & 0 \\ 0 & 1 & 0 & 0 \end{bmatrix}^2 = \begin{bmatrix} 1 & 0 & 0 & 0 \\ 0 & 1 & 0 & 0 \\ 0 & 0 & 1 & 0 \\ 0 & 0 & 0 & 1 \end{bmatrix} \qquad (4.8)$$

We see that symmetric permutation matrices are square roots of unity[15].

Cyclic groups of permutation matrices:
The powers of any permutation matrix form a set of permutation matrices which are a multiplicatively closed set. Such a closed set includes the identity matrix, of course, and also includes the inverses of each matrix within the set. If a permutation matrix, P, is such that:

$$P^n = I \qquad (4.9)$$

wherein we have used the I to denote the identity matrix, then:

$$P^{(n-m)} P^m = I \qquad (4.10)$$

We see that P^m is the inverse of $P^{(n-m)}$.

Consider the matrices:

---

[14] The permutation matrix with 1's on the diagonal from top right to bottom left is sometimes called the counter identity matrix. Perhaps this is not sensible nomenclature.
[15] Symmetric matrices have the 'Hermitian' property of having real eigenvalues and orthogonal eigenvectors.

The Order of a Permutation Matrices

$$\begin{bmatrix} 0 & 1 & 0 & 0 \\ 0 & 0 & 1 & 0 \\ 0 & 0 & 0 & 1 \\ 1 & 0 & 0 & 0 \end{bmatrix}^2 = \begin{bmatrix} 0 & 0 & 1 & 0 \\ 0 & 0 & 0 & 1 \\ 1 & 0 & 0 & 0 \\ 0 & 1 & 0 & 0 \end{bmatrix}$$

$$\begin{bmatrix} 0 & 1 & 0 & 0 \\ 0 & 0 & 1 & 0 \\ 0 & 0 & 0 & 1 \\ 1 & 0 & 0 & 0 \end{bmatrix}^3 = \begin{bmatrix} 0 & 0 & 0 & 1 \\ 1 & 0 & 0 & 0 \\ 0 & 1 & 0 & 0 \\ 0 & 0 & 1 & 0 \end{bmatrix} \qquad \begin{bmatrix} 0 & 1 & 0 & 0 \\ 0 & 0 & 1 & 0 \\ 0 & 0 & 0 & 1 \\ 1 & 0 & 0 & 0 \end{bmatrix}^4 = \begin{bmatrix} 1 & 0 & 0 & 0 \\ 0 & 1 & 0 & 0 \\ 0 & 0 & 1 & 0 \\ 0 & 0 & 0 & 1 \end{bmatrix}$$

(4.11)

Since raising the permutation matrix in (4.11) to power five is just multiplying the matrix by the identity, we see that the cycle of permutation matrices will repeat itself. We reiterate, the set of permutation matrices which are the powers of a particular permutation matrix are a closed set of permutation matrices which includes the identity matrix.

Cyclic groups:
Such a set of permutation matrices which are generated by a single permutation matrix as the powers of that permutation matrix is called a cyclic group of permutation matrices.

The cyclic group of three permutation matrices is:

$$C_3 = \left\{ \begin{bmatrix} 1 & 0 & 0 \\ 0 & 1 & 0 \\ 0 & 0 & 1 \end{bmatrix}, \begin{bmatrix} 0 & 1 & 0 \\ 0 & 0 & 1 \\ 1 & 0 & 0 \end{bmatrix}, \begin{bmatrix} 0 & 0 & 1 \\ 1 & 0 & 0 \\ 0 & 1 & 0 \end{bmatrix} \right\} \qquad (4.12)$$

We have used the symbol C_3 to denote the cyclic group of three matrices. This is standard notation. The point to notice is that each of the two non-identity 3×3 matrices in (4.12) will generate this cyclic group. If we take the powers of either of the two non-identity matrices in (4.12), we get the same set of matrices, (4.12).

Consider the six 3×3 permutation matrices; this is the complete set of 3×3 permutation matrices:

$$\begin{bmatrix} 1 & 0 & 0 \\ 0 & 1 & 0 \\ 0 & 0 & 1 \end{bmatrix} \begin{bmatrix} 0 & 1 & 0 \\ 0 & 0 & 1 \\ 1 & 0 & 0 \end{bmatrix} \begin{bmatrix} 0 & 0 & 1 \\ 1 & 0 & 0 \\ 0 & 1 & 0 \end{bmatrix}$$
$$\begin{bmatrix} 1 & 0 & 0 \\ 0 & 0 & 1 \\ 0 & 1 & 0 \end{bmatrix} \begin{bmatrix} 0 & 0 & 1 \\ 0 & 1 & 0 \\ 1 & 0 & 0 \end{bmatrix} \begin{bmatrix} 0 & 1 & 0 \\ 1 & 0 & 0 \\ 0 & 0 & 1 \end{bmatrix} \qquad (4.13)$$

With calculation, we find that one of these six permutation matrices, the identity, is of order one, three of these six permutation matrices, the symmetric ones, are of order two, and two of these six permutation matrices are of order three. It would seem that we have one cyclic group of one matrix – correct, this is the identity matrix. It would seem that we have three cyclic groups of two matrices – correct, each of these is a different symmetric matrix and the identity. It would seem that we have two cyclic groups of three matrices – not correct; the two sets of three matrices are identical; we have one cyclic group of three matrices.

Symmetric groups:
Any complete set of every $n\times n$ permutation matrix is called a symmetric group. The complete set of 3×3 permutation matrices shown in (4.13) is the symmetric group S_3 - the 3 designates the size of the matrices which is the same as the number of objects being permuted. Such a set is closed under matrix multiplication, contains inverses of every element in the set, and includes the identity matrix. However, such a complete set cannot be generated as the powers of a single permutation matrix. Looking at the complete set, symmetric group S_3, (4.13), and the cyclic group C_3, (4.12), we see that the three permutation matrices in (4.12) could never generate the whole of S_3. Furthermore, no other single member of the six S_3 matrices, (4.13), could generate the whole group as powers of itself.

More than one generating element:

Suppose we multiply two of the permutation matrices in (4.13) together and take the various powers of these two permutation matrices and multiply these various powers together; would this produce the whole symmetric group? Yes it would if we are careful to pick the correct pair of permutation matrices. Choose:

$$\begin{bmatrix} 0 & 1 & 0 \\ 0 & 0 & 1 \\ 1 & 0 & 0 \end{bmatrix} \& \begin{bmatrix} 1 & 0 & 0 \\ 0 & 0 & 1 \\ 0 & 1 & 0 \end{bmatrix} \quad (4.14)$$

These two matrices, (4.14), together will generate the whole symmetric group. They are not the only pair of matrices that will generate the whole symmetric group.

Summary:

The order of a permutation matrix is the power to which that matrix must be raised to produce the identity matrix.

The set of m permutation matrices generated as the powers of a single permutation matrix, including the identity matrix, is called a cyclic group and is denoted by C_m.

The set of all $n!$ permutation matrices of size $n \times n$ is called a symmetric group and is denoted by S_n. In general, for $n > 2$, the whole of the group S_n cannot be generated by any single element of S_n.

Chapter 5

Finite groups and Representations

The reader who has previously studied finite group theory might, by now, have realised that the mathematics of permutation matrices, the mathematics of permutations, is finite group theory. If you have not previously studied finite group theory, be appraised that you are now studying finite group theory and you have been doing so since the beginning of this book – isn't it easy. We are often told that finite group theory is the mathematics of symmetry. Your author is disinclined from this view; we take the view that finite group theory is no more than the mathematics of permutations. This is a simpler view[16].

Finite group definition:

A finite group is nothing more than a set of permutations (think permutation matrices) which:

a) Are closed under sequential combination of permutations (matrix multiplication).
b) Include the identity permutation (identity matrix).
c) Include an inverse of every element in the set.
d) Are associative.

It is the nature of the combination of permutations, and of the multiplication of matrices, that these operations are associative, and so we need not consider the associativity requirement any further.

---

[16] To have a symmetry, we have to assume a particular type of space upon which to impose finite groups. Finite groups are more fundamental than that; we can add a type of space in later if we wish – we won't so wish in this book.

Finite Groups and Representations

The order of a finite group:

The number of permutations in a given finite group is called the order of that group. This is a concept different from the order of an element of the group. Sorry, there are too many different uses of the word order in this area of mathematics.

Seeking finite groups:

It is one of the pre-occupations of mathematicians that they seek sets of permutation matrices, sets of permutations, which form finite groups.

For a start, the whole set of permutations of n objects, which is equivalent to the whole set of $n \times n$ permutation matrices, is always a finite group; it is called a symmetric group and is denoted by S_n. There are $n!$ elements in the finite group S_n. Also, the whole set of even permutations of n objects ($n \times n$ permutation matrices with determinant $+1$) is always a finite group; it is called an alternating group and is denoted by A_n. There are $\dfrac{n!}{2}$ elements in the alternating finite group A_n. We have:

$$S_n = \begin{bmatrix} All\ n \times n \\ Perm\ Mats \end{bmatrix} \qquad A_n = \begin{bmatrix} All\ n \times n \\ Perm\ Mats \\ with\ Det = +1 \end{bmatrix} \quad (5.1)$$

$$\text{order } n! \qquad\qquad \text{order } \dfrac{n!}{2}$$

An alternating group, A_n, can always be written using $\dfrac{n}{2} \times \dfrac{n}{2}$ permutation matrices.

That bit was easy, but there are other types of finite groups. Finding the other finite groups is not so easy.

The order 1 finite group:
We call the number of elements, permutations or permutation matrices, in a finite group the order of the group. There is only one finite group of order 1. Since every finite group must have an identity permutation, the order 1 finite group is just the identity. As permutation matrices, we have:

$$[1], \begin{bmatrix} 1 & 0 \\ 0 & 1 \end{bmatrix}, \begin{bmatrix} 1 & 0 & 0 \\ 0 & 1 & 0 \\ 0 & 0 & 1 \end{bmatrix}, \quad \ldots \quad (5.2)$$

Each of these permutation matrices, (5.2), is a way of writing, a representation of, the single order 1 finite group which is the identity.

Representations of finite groups:
Every finite group can be written as a set of $n \times n$ permutation matrices, but, more than this, there are several sets of different sized permutation matrices which represent the same finite group. The example of the order 1 finite group above, (5.2), is just one example of this phenomenon. We call each way of writing a finite group, each different size of permutation matrices used to represent that finite group, a representation of that group. Associated with a given finite group are the many representations (sets of different sized permutation matrices) of that group. Of course, each set has the same number of matrices – it is only the size of the matrices that differs.

The order 2 finite group:
There is only one order 2 finite group. Since there are two elements in the order 2 finite group, it cannot be represented by only one permutation matrix. Since there is only one 1×1 permutation matrix, the order 2 finite group cannot be represented by 1×1 permutation matrices.

There are two 2×2 permutation matrices. We use the 2×2 permutation matrices to represent the order 2 finite group:

Finite Groups and Representations

$$\begin{bmatrix} 1 & 0 \\ 0 & 1 \end{bmatrix} \quad \& \quad \begin{bmatrix} 0 & 1 \\ 1 & 0 \end{bmatrix} \tag{5.3}$$

We see that we have the identity matrix. We see that each of these matrices, (5.3), is equal to its own transpose, and so we have an inverse of each element – each element is its own inverse. This set of permutation matrices is closed under sequential combination of permutations, matrix multiplication, because all complete sets of $n \times n$ permutation matrices are closed.

Note that we have one even permutation matrix and one odd permutation matrix in (5.3).

The order two group is most often called the cyclic group of order two, and it is denoted by C_2. It is, of course, also the symmetric group S_2. Two names, same thing.

Aside:
The reader might come across the order two finite group being represented by the two 1×1 matrices:

$$[1] \quad \& \quad [-1] \tag{5.4}$$

The claim is made that this is a 1-dimensional representation of the order two group because it is formed from 1×1 matrices. We dislike using minus signs to distort the picture, and so we will not proceed along this path.

Similarly, the order three group is sometimes represented by the three 1×1 matrices:

$$[1] \quad \& \quad \left[-\frac{1}{2} + i\frac{\sqrt{3}}{2} \right] \quad \& \quad \left[-\frac{1}{2} - i\frac{\sqrt{3}}{2} \right] \tag{5.5}$$

These can be written as the 2×2 matrices:

$$\begin{bmatrix} 1 & 0 \\ 0 & 1 \end{bmatrix} \quad \& \quad \begin{bmatrix} -\frac{1}{2} & \frac{\sqrt{3}}{2} \\ -\frac{\sqrt{3}}{2} & -\frac{1}{2} \end{bmatrix} \quad \& \quad \begin{bmatrix} -\frac{1}{2} & -\frac{\sqrt{3}}{2} \\ \frac{\sqrt{3}}{2} & -\frac{1}{2} \end{bmatrix} \quad (5.6)$$

We do not enjoy using either minus signs or higher dimensional complex numbers to 'distort' the simplicity of the representations of finite groups. When we speak of representations, we mean only matrices with 1's as elements.

The order 3 finite group:
There are six 3×3 permutation matrices:

$$\begin{bmatrix} 1 & 0 & 0 \\ 0 & 1 & 0 \\ 0 & 0 & 1 \end{bmatrix}, \begin{bmatrix} 0 & 1 & 0 \\ 0 & 0 & 1 \\ 1 & 0 & 0 \end{bmatrix}, \begin{bmatrix} 0 & 0 & 1 \\ 1 & 0 & 0 \\ 0 & 1 & 0 \end{bmatrix}$$
$$\begin{bmatrix} 1 & 0 & 0 \\ 0 & 0 & 1 \\ 0 & 1 & 0 \end{bmatrix}, \begin{bmatrix} 0 & 0 & 1 \\ 0 & 1 & 0 \\ 1 & 0 & 0 \end{bmatrix}, \begin{bmatrix} 0 & 1 & 0 \\ 1 & 0 & 0 \\ 0 & 0 & 1 \end{bmatrix} \quad (5.7)$$

If we are to form an order 3 finite group from three of these six matrices, (5.7), then we must include the identity, but which other two of these six matrices must we include? This question is easily answered in the case of the 3×3 permutation matrices by brute force calculation; we simply multiply pairs of matrices together looking for multiplicative closure, inverses etc. However, if we ask the same question of sets of larger permutation matrices things get a little more difficult. There are twenty-four 4×4 permutation matrices, and there are one hundred and twenty 5×5 permutation matrices. We see that, although in every case brute force calculation would solve the problem, the amount of brute force needed increases dramatically as we increase the size of the permutation matrices.

Finite Groups and Representations

One of the questions in mathematics is how do we efficiently choose all sets of matrices which form a finite group from the full set of $n!$ $n \times n$ permutation matrices?

The C_2 subgroups of S_3:
Looking at the matrices in (5.7), we see that the three matrices on the bottom row are all equal to their transpose – they are symmetric across the leading diagonal. This means that they are all their own inverse[17]. A matrix which is its own inverse, together with the identity, forms the order two group, C_2. Thus, we have three copies of the order two group, C_2, within the six 3×3 permutation matrices above, (5.7); these are:

$$\left\{ \begin{bmatrix} 1 & 0 & 0 \\ 0 & 1 & 0 \\ 0 & 0 & 1 \end{bmatrix}, \begin{bmatrix} 1 & 0 & 0 \\ 0 & 0 & 1 \\ 0 & 1 & 0 \end{bmatrix} \right\} \quad \left\{ \begin{bmatrix} 1 & 0 & 0 \\ 0 & 1 & 0 \\ 0 & 0 & 1 \end{bmatrix}, \begin{bmatrix} 0 & 1 & 0 \\ 1 & 0 & 0 \\ 0 & 0 & 1 \end{bmatrix} \right\}$$

$$\left\{ \begin{bmatrix} 1 & 0 & 0 \\ 0 & 1 & 0 \\ 0 & 0 & 1 \end{bmatrix}, \begin{bmatrix} 0 & 0 & 1 \\ 0 & 1 & 0 \\ 1 & 0 & 0 \end{bmatrix} \right\}$$

(5.8)

These, (5.8), are the three 3-dimensional representations of the order two group.

Aside:
Since any symmetric matrix always forms the finite group C_2 when put together with the identity matrix, and since the representation of C_2 given above, (5.3), is one even permutation matrix, the identity, and one odd permutation matrix, we might conclude that all symmetric permutation matrices are odd permutation matrices. We would be wrong. A counter-example is the symmetric matrix:

---

[17] Careful; this is the case for only permutation matrices.

$$\det\begin{pmatrix}\begin{bmatrix}0 & 1 & 0 & 0\\ 1 & 0 & 0 & 0\\ 0 & 0 & 0 & 1\\ 0 & 0 & 1 & 0\end{bmatrix}\end{pmatrix} = +1 \tag{5.9}$$

This, (5.9), is an even permutation matrix. Together with the 4×4 identity matrix, this matrix, (5.9), is a representation the group C_2.

The conclusion is that the oddness/evenness of the elements of a finite group vary with the representation of that finite group and are not an invariant property of a finite group that transcends different representations. We cannot say that the finite group C_2 is comprised of one odd permutation matrix and one even permutation matrix. We can say that the 2-dimensional representation (2×2 permutation matrices) of the finite group C_2 is comprised of one odd permutation matrix and one even permutation matrix.

In due course, we will find that each finite group has a 'special' representation called the adjoint representation of that group. We can always say that the adjoint representation of a particular finite group is comprised of a definite number of odd permutation matrices and a definite number of even permutation matrices.

Back to the order three group:
With thought, we see that the three matrices on the top row of (5.7) form an order three group; this group is the order three cyclic group denoted by C_3. We note that these three matrices are all even matrices in this representation.

Back to the symmetric groups:
Every complete set of $n\times n$ permutation matrices includes the identity matrix, includes the inverses of every $n\times n$ permutation matrix, and is

closed under matrix multiplication (sequential combination of permutations).

Complete sets of permutations must be this way; think back to the swapping of the positions of objects; if we have a complete set of these swaps, then we have every inverse[18], we have closure, and we have the identity.

These complete sets of $n \times n$ permutation matrices are called the symmetric groups. The symmetric group which is the permutations of n objects represented by $n!$ $n \times n$ permutation matrices is denoted by S_n. Examples of symmetric groups include S_1 which is just the identity matrix, (5.2), and include S_2 which is the two 2×2 permutation matrices, (5.3), and include S_3 which is the six 3×3 permutation matrices above, (5.7). We have:

$$S_3 \equiv \left\{ \begin{bmatrix} 1 & 0 & 0 \\ 0 & 1 & 0 \\ 0 & 0 & 1 \end{bmatrix}, \begin{bmatrix} 0 & 1 & 0 \\ 0 & 0 & 1 \\ 1 & 0 & 0 \end{bmatrix}, \begin{bmatrix} 0 & 0 & 1 \\ 1 & 0 & 0 \\ 0 & 1 & 0 \end{bmatrix}, \begin{bmatrix} 1 & 0 & 0 \\ 0 & 0 & 1 \\ 0 & 1 & 0 \end{bmatrix}, \begin{bmatrix} 0 & 0 & 1 \\ 0 & 1 & 0 \\ 1 & 0 & 0 \end{bmatrix}, \begin{bmatrix} 0 & 1 & 0 \\ 1 & 0 & 0 \\ 0 & 0 & 1 \end{bmatrix} \right\} \quad (5.10)$$

Groups other than the symmetric groups:
Looking at the six 3×3 permutation matrices above, (5.7), we have seen that the three matrices on the top row are a finite group, C_3:

$$C_3 \equiv \left\{ \begin{bmatrix} 1 & 0 & 0 \\ 0 & 1 & 0 \\ 0 & 0 & 1 \end{bmatrix}, \begin{bmatrix} 0 & 1 & 0 \\ 0 & 0 & 1 \\ 1 & 0 & 0 \end{bmatrix}, \begin{bmatrix} 0 & 0 & 1 \\ 1 & 0 & 0 \\ 0 & 1 & 0 \end{bmatrix} \right\} \quad (5.11)$$

---

[18] Simply reverse the permutation.

These three matrices are all even permutation matrices in this representation. Although this, (5.11), is a cyclic group, C_3, it is also an alternating group and could equally well have been denoted by A_3. Similarly, the symmetric group, S_2, above, (5.3), is also a cyclic group and can be denoted by C'_2.

All finite groups are subgroups of symmetric groups:
It is established mathematical fact that every finite group is a subgroup of some symmetric group. Since every complete set of permutation matrices is a symmetric group, and since every finite group is a set of permutation matrices, it is clear that every finite group is a subgroup of some symmetric group. This does not mean that a particular finite group is a subgroup of a particular symmetric group; for example the order four cyclic group C_4 is not a subgroup of the order six symmetric group S_3 - we cannot find four of the six 3×3 permutation matrices, (5.10), which form an order four group. However, the order four cyclic group C_4 is a subgroup of the order twenty-four symmetric group S_4.

Considering what we have discovered earlier in this chapter, and with the above six matrices, (5.10), in mind, we can see that the order six symmetric group S_3 has three C_2 subgroups and one C_3 subgroup. As with all finite groups, the identity is an order one subgroup and the entire group S_3 can also be thought of as a subgroup of itself. We now have the entire subgroup structure of S_3.

Symmetric groups as subgroups:
Every symmetric group S_n is a subgroup of the next higher order symmetric group S_{n+1}. This is most easy to see in the standard two line permutation notation. We can write an order two permutation as an order three permutation with one object not changing position. In the case of the $S_2 = C_2$ symmetric group of order two, we have three ways

of doing this each of which corresponds to holding constant each of the three numbers in the top row of the permutation:

$$C_2 \equiv \begin{pmatrix} 1 & 2 & 3 \\ 1 & 2 & 3 \end{pmatrix} \& \begin{pmatrix} 1 & 2 & 3 \\ 1 & 3 & 2 \end{pmatrix} \quad C_2 \equiv \begin{pmatrix} 1 & 2 & 3 \\ 1 & 2 & 3 \end{pmatrix} \& \begin{pmatrix} 1 & 2 & 3 \\ 3 & 2 & 1 \end{pmatrix}$$

$$C_2 \equiv \begin{pmatrix} 1 & 2 & 3 \\ 1 & 2 & 3 \end{pmatrix} \& \begin{pmatrix} 1 & 2 & 3 \\ 2 & 1 & 3 \end{pmatrix}$$

(5.12)

The three non-identity permutations in (5.12) correspond to the three permutation matrices on the bottom row of (5.10). Holding one position unchanged in a two line permutation corresponds to having a 1 on the leading diagonal in the corresponding permutation matrix:

$$\begin{pmatrix} 1 & 2 & 3 \\ 1 & \sim & \sim \end{pmatrix} \rightarrow \begin{bmatrix} 1 & 0 & 0 \\ 0 & \sim & \sim \\ 0 & \sim & \sim \end{bmatrix} \qquad (5.13)$$

We see that there are three ways to form the order two symmetric/cyclic group $S_2 = C_2$ from the set of six order three permutations. This means there are three copies of the order two symmetric group $S_2 = C_2$ in the order six symmetric group S_3 [19].

With a little thought, we see that there will be four copies of the order six symmetric group S_3 in the order twenty-four symmetric group S_4 because we can take the six order three permutations, (5.10), in S_3 and add an extra object which does not change position. This is equivalent to expanding the 3×3 permutation matrices to be 4×4 permutation matrices with a 1 on the leading diagonal:

---

[19] The order six symmetric group S_3 is also known as the order six dihedral group D_3.

$$\begin{pmatrix} 1 & 2 & 3 \\ 1 & 3 & 2 \end{pmatrix} \equiv \begin{bmatrix} 1 & 0 & 0 \\ 0 & 0 & 1 \\ 0 & 1 & 0 \end{bmatrix} \rightarrow \begin{pmatrix} 1 & 2 & 3 & 4 \\ 1 & 3 & 2 & 4 \end{pmatrix} \equiv \begin{bmatrix} 1 & 0 & 0 & 0 \\ 0 & 0 & 1 & 0 \\ 0 & 1 & 0 & 0 \\ 0 & 0 & 0 & 1 \end{bmatrix}$$

(5.14)

In general, we see that the symmetric group $S_n : n > 2$ will have n copies of $S_{(n-1)}$ as subgroups.

Further, since there are six ways of holding two of four positions unchanged, there will be six order two $S_2 \equiv C_2$ subgroups in S_4. Since there are four S_3 subgroups in S_4 and since there are three S_2 subgroups in S_3, we might naively expect there would be $4 \times 3 = 12$ S_2 subgroups in S_4. It is not so, some of the S_2 subgroups are 'shared' among the S_3 subgroups.

Summary:
Permutation matrices are a refreshingly different way of doing finite group theory.

Chapter 6

Special Representations of Finite Groups

Although a particular finite group can be represented in many ways by many sets of differently sized permutation matrices, there are two 'special' particular sizes of permutation matrices which are associated with a given group. These two sizes of permutation matrices representing a given group are called the fundamental representation of that group and the adjoint representation of that group.

The fundamental representation of a group:
The fundamental representation of a group is that representation of the group as $n \times n$ permutation matrices where n is the smallest size of permutation matrices[20] which can represent the group.

The fundamental representation of S_3 is the six 3×3 matrices given above, (5.10). The fundamental representation of C_3 is the three 3×3 matrices given above, (5.11). The fundamental representation of C_2 is the three 2×2 matrices given above, (5.3).

The fundamental representation is not necessarily unique except for the symmetric groups and the alternating groups. There are three fundamental representations of C_4 which are the three sets of four 4×4 matrices given below, (6.5) & (6.6).

In general, the fundamental representation of the symmetric group S_n is the $n!$ $n \times n$ permutation matrices. In general, the fundamental

---

[20] This is permutation matrices and not matrices in general.

representation of the alternating group A_n is the $\frac{n!}{2}$ even $n \times n$ permutation matrices.

The adjoint representation of a group:
The adjoint representation of a group is the representation of that group which uses permutation matrices of the same size as the order of the group – remember, the order of a group is the number of elements in the group.

Above, (5.10) we have presented the order six symmetric group S_3 as six 3×3 permutation matrices; we could have presented this same group as six 6×6 permutation matrices:

$$S_3 = \begin{bmatrix} a & b & c & d & e & f \\ c & a & b & e & f & d \\ b & c & a & f & d & e \\ d & e & f & a & b & c \\ e & f & d & c & a & b \\ f & d & e & b & c & a \end{bmatrix} \qquad (6.1)$$

In this, (6.1), we have identified each separate permutation matrix with a different variable; for example:

$$\begin{bmatrix} 0 & 0 & 0 & 0 & 0 & f \\ 0 & 0 & 0 & 0 & f & 0 \\ 0 & 0 & 0 & f & 0 & 0 \\ 0 & 0 & f & 0 & 0 & 0 \\ 0 & f & 0 & 0 & 0 & 0 \\ f & 0 & 0 & 0 & 0 & 0 \end{bmatrix} \rightarrow \begin{bmatrix} 0 & 0 & 0 & 0 & 0 & 1 \\ 0 & 0 & 0 & 0 & 1 & 0 \\ 0 & 0 & 0 & 1 & 0 & 0 \\ 0 & 0 & 1 & 0 & 0 & 0 \\ 0 & 1 & 0 & 0 & 0 & 0 \\ 1 & 0 & 0 & 0 & 0 & 0 \end{bmatrix} \qquad (6.2)$$

This representation of the symmetric group S_3 (the six permutation matrices) is called the adjoint representation of S_3.

The adjoint representation of the cyclic group C_3 is:

$$\begin{bmatrix} a & b & c \\ c & a & b \\ b & c & a \end{bmatrix} \equiv \left\{ \begin{bmatrix} 1 & 0 & 0 \\ 0 & 1 & 0 \\ 0 & 0 & 1 \end{bmatrix}, \begin{bmatrix} 0 & 1 & 0 \\ 0 & 0 & 1 \\ 1 & 0 & 0 \end{bmatrix}, \begin{bmatrix} 0 & 0 & 1 \\ 1 & 0 & 0 \\ 0 & 1 & 0 \end{bmatrix} \right\} \quad (6.3)$$

For cyclic groups, the adjoint representation is the same size, and the same permutation matrices, as the fundamental representation.

The adjoint representation of the order twenty-four symmetric group, S_4, is twenty-four 24×24 matrices. The adjoint representation of the order twelve alternating group, A_4, is twelve 12×12 matrices.

The adjoint representations are not unique. Below, (6.5) & (6.6), we have the three different adjoint representations of the cyclic group C_4. Although there is most often several adjoint representations of a particular group, the properties associated with the adjoint representation such as the oddness and evenness of the constituent permutation matrices is the same for all adjoint representations of the same group.

A peek ahead:
We are most interested in the adjoint representations of the finite groups because taking the matrix exponential of an adjoint representation produces a type of numbers – a type of division algebra like the complex numbers or the quaternions. The exponential produces the polar form of the division algebra which is comprised of a rotation matrix (the imaginary variables) containing trigonometric functions and a radial matrix (the real variable). We have a peek ahead:

$$\exp(C_3^{Adjoint}) = \exp\left(\begin{bmatrix} a & b & c \\ c & a & b \\ b & c & a \end{bmatrix}\right)$$

$$= \begin{bmatrix} r & 0 & 0 \\ 0 & r & 0 \\ 0 & 0 & r \end{bmatrix} \begin{bmatrix} v_A(b,c) & v_B(b,c) & v_C(b,c) \\ v_C(b,c) & v_A(b,c) & v_B(b,c) \\ v_B(b,c) & v_C(b,c) & v_A(b,c) \end{bmatrix} \quad (6.4)$$

In this, (6.4), the $v_i(b,c)$ are the 3-dimensional trigonometric functions and the matrix containing them is the 3-dimensional rotation matrix of this space.[21]

Adjoint representations and Cayley tables:
Associated with every finite group is a multiplication table which presents the products of the elements of the finite group. Such a multiplication table is called a Cayley table of a particular finite group. The Cayley table of a group encodes the multiplicative structure of the group; different groups have different multiplicative structures. An example of a Cayley table is the Cayley table of the C_3 group:

	A	B	C
A	a	b	c
B	b	c	a
C	c	a	b

In this Cayley table, we see that the combination of the two A permutations (A is the identity) is the a (identity) permutation and the combination of the two B permutations is the c permutation. Since the Cayley table is just a multiplication table, we can swap rows or columns as we choose. We will swap the columns to give the identity, a, on the leading diagonal.

---

[21] See: Dennis Morris : Complex Numbers The Higher Dimensional Forms.

We do not need the topmost line of the above Cayley table nor the left-most line of the above Cayley table as these are just repeated in the topmost row and the left-most column of the body of the table. Swapping some columns and renaming some variables, we can write our Cayley table as:

a	b	c
c	a	b
b	c	a

A Cayley table written with all the identities on the leading diagonal is called a Standard Form Cayley table. All groups have at least one Standard Form Cayley table. Most groups have more than one Standard Form Cayley table. We can obtain these different forms of Standard Form Cayley tables from each other by simply swapping the variable names and then swapping rows around.

We will now use matrix notation to present the standard form Cayley tables. The cyclic group C_4 has three different forms of its Standard Form Cayley table:

$$C_4 = \begin{bmatrix} a & b & c & d \\ d & a & b & c \\ c & d & a & b \\ b & c & d & a \end{bmatrix} \tag{6.5}$$

$$C_4 = \begin{bmatrix} a & b & c & d \\ b & a & d & c \\ d & c & a & b \\ c & d & b & a \end{bmatrix} \quad \& \quad C_4 = \begin{bmatrix} a & b & c & d \\ c & a & d & b \\ b & d & a & c \\ d & c & b & a \end{bmatrix} \tag{6.6}$$

The adjoint representation of a group is a direct copy of the Standard Form Cayley table of the group. We have presented the above, (6.5) & (6.6), the three Standard form Cayley tables of the group C_4; these are also the three adjoint representations of the group C_4. As we see above, (6.5) & (6.6), groups can have more than one adjoint representation.

The different adjoint representations are essentially the same, isomorphic, just as the different Standard Form Cayley tables are essentially the same.

Summing permutation matrices:

Every finite group has a Cayley table. Every Cayley table can be presented in Standard Form (with all the identities on the leading diagonal). Therefore every finite group has an adjoint representation.

The nature of all Cayley tables is that there is one, and only one, element in every position in the table. The nature of Cayley tables is that a particular element occurs once, and only once, in each row of the table and in each column of the table.

We can separate the parts of the adjoint representation of a group which is the Standard form Cayley table of that group. In the case of the C_4 group adjoint representation above, (6.5), we get:

$$\begin{bmatrix} a & 0 & 0 & 0 \\ 0 & a & 0 & 0 \\ 0 & 0 & a & 0 \\ 0 & 0 & 0 & a \end{bmatrix} \& \begin{bmatrix} 0 & b & 0 & 0 \\ 0 & 0 & b & 0 \\ 0 & 0 & 0 & b \\ b & 0 & 0 & 0 \end{bmatrix}$$
$$\begin{bmatrix} 0 & 0 & c & 0 \\ 0 & 0 & 0 & c \\ c & 0 & 0 & 0 \\ 0 & c & 0 & 0 \end{bmatrix} \& \begin{bmatrix} 0 & 0 & 0 & d \\ d & 0 & 0 & 0 \\ 0 & d & 0 & 0 \\ 0 & 0 & d & 0 \end{bmatrix}$$
(6.7)

Of course, we used the different variables to replace the 1's. We see that the adjoint representation of a group is a set of permutation matrices which add to an 'all 1's' matrix:

Special Representations of Finite Groups

$$\begin{bmatrix} 1 & 0 & 0 & 0 \\ 0 & 1 & 0 & 0 \\ 0 & 0 & 1 & 0 \\ 0 & 0 & 0 & 1 \end{bmatrix} + \begin{bmatrix} 0 & 1 & 0 & 0 \\ 0 & 0 & 1 & 0 \\ 0 & 0 & 0 & 1 \\ 1 & 0 & 0 & 0 \end{bmatrix} + \begin{bmatrix} 0 & 0 & 1 & 0 \\ 0 & 0 & 0 & 1 \\ 1 & 0 & 0 & 0 \\ 0 & 1 & 0 & 0 \end{bmatrix} + \begin{bmatrix} 0 & 0 & 0 & 1 \\ 1 & 0 & 0 & 0 \\ 0 & 1 & 0 & 0 \\ 0 & 0 & 1 & 0 \end{bmatrix}$$

$$= \begin{bmatrix} 1 & 1 & 1 & 1 \\ 1 & 1 & 1 & 1 \\ 1 & 1 & 1 & 1 \\ 1 & 1 & 1 & 1 \end{bmatrix}$$

(6.8)

This is an important realisation. If we use the adjoint representation of a group, then, we can search for groups of a particular order by searching through all the permutation matrices of that particular size looking for sets of permutation matrices which add to form an 'all 1's' matrix.

Warning:

Be warned, not all such 'all 1's' sums of permutation matrices are groups – see next chapter – the key is that the identity matrix must be one of the permutation matrices which are added to form the 'all 1's' matrix.

Chapter 7

Casting out the Order Four Groups

We are going to find all the order four groups by combing through all the 4×4 permutation matrices looking for sets of permutation matrices which sum to an 'all 1's' matrix. We are using the fact that the adjoint representation of any finite group will be a set of permutation matrices which, when the permutation matrices are summed, will produce an 'all 1's' matrix.

Every finite group must have an identity. The 4×4 identity matrix is:

$$\begin{bmatrix} 1 & 0 & 0 & 0 \\ 0 & 1 & 0 & 0 \\ 0 & 0 & 1 & 0 \\ 0 & 0 & 0 & 1 \end{bmatrix} \qquad (7.1)$$

Since we seek 'all 1's' matrices, we can immediately cast aside any 4×4 permutation matrix which has a 1, or more than one 1, on the leading diagonal; these permutation matrices can never be added to the identity matrix to form an 'all 1's' matrix, and, without the identity, these permutation matrices cannot be an element of any group with an adjoint representation of the size we seek[22].

We can further cast aside any permutation matrix that is such that any power of it has a 1 on the leading diagonal unless this power is the identity matrix. Remember, a finite group contains all the powers of each permutation matrix within it; if one of these powers has a single 1 on the leading diagonal, then we would not have an 'all 1's' matrix when the identity was added. In the case of the 4×4 permutation

---

[22] They could be part of a representation of a group other than the adjoint, and all permutation matrices are elements of some symmetric group.

matrices, this is not a problem because there are no such 4×4 permutation matrices. There are such larger permutation matrices.

There are $4!=24$ 4×4 permutation matrices. Casting aside the matrices which do not have all zeros on the leading diagonal, we are left with:

$$b: \begin{bmatrix} 0 & 1 & 0 & 0 \\ 1 & 0 & 0 & 0 \\ 0 & 0 & 0 & 1 \\ 0 & 0 & 1 & 0 \end{bmatrix}, \begin{bmatrix} 0 & 1 & 0 & 0 \\ 0 & 0 & 1 & 0 \\ 0 & 0 & 0 & 1 \\ 1 & 0 & 0 & 0 \end{bmatrix}, \begin{bmatrix} 0 & 1 & 0 & 0 \\ 0 & 0 & 0 & 1 \\ 1 & 0 & 0 & 0 \\ 0 & 0 & 1 & 0 \end{bmatrix} \quad (7.2)$$

$$c: \begin{bmatrix} 0 & 0 & 1 & 0 \\ 1 & 0 & 0 & 0 \\ 0 & 0 & 0 & 1 \\ 0 & 1 & 0 & 0 \end{bmatrix}, \begin{bmatrix} 0 & 0 & 1 & 0 \\ 0 & 0 & 0 & 1 \\ 1 & 0 & 0 & 0 \\ 0 & 1 & 0 & 0 \end{bmatrix}, \begin{bmatrix} 0 & 0 & 1 & 0 \\ 0 & 0 & 0 & 1 \\ 0 & 1 & 0 & 0 \\ 1 & 0 & 0 & 0 \end{bmatrix} \quad (7.3)$$

$$d: \begin{bmatrix} 0 & 0 & 0 & 1 \\ 1 & 0 & 0 & 0 \\ 0 & 1 & 0 & 0 \\ 0 & 0 & 1 & 0 \end{bmatrix}, \begin{bmatrix} 0 & 0 & 0 & 1 \\ 0 & 0 & 1 & 0 \\ 1 & 0 & 0 & 0 \\ 0 & 1 & 0 & 0 \end{bmatrix}, \begin{bmatrix} 0 & 0 & 0 & 1 \\ 0 & 0 & 1 & 0 \\ 0 & 1 & 0 & 0 \\ 1 & 0 & 0 & 0 \end{bmatrix} \quad (7.4)$$

The above nine matrices, (7.2) & (7.3) & (7.4), need to be combined (added) to the identity matrix to form an 'all 1's' matrix. Since there are nine above matrices, we might think this means there are only three order four groups which can be formed from the 4×4 permutation matrices because there can be only three sets of three matrices; this would be a mistake; things are not that simple.

Looking at the top rows of the above nine all zeros matrices, we see that we must form combinations which include one from each of the three sets of matrices labelled $\{b,c,d\}$.

The b matrix:

$$\begin{bmatrix} 0 & 1 & 0 & 0 \\ 1 & 0 & 0 & 0 \\ 0 & 0 & 0 & 1 \\ 0 & 0 & 1 & 0 \end{bmatrix} \tag{7.5}$$

cannot combine with any other b matrix to form an 'all 1's' matrix because of the coincident 1's on the top row of these matrices; nor can it combine with the particular c matrix or the particular d matrix:

$$c: \begin{bmatrix} 0 & 0 & 1 & 0 \\ 1 & 0 & 0 & 0 \\ 0 & 0 & 0 & 1 \\ 0 & 1 & 0 & 0 \end{bmatrix} \quad \& \quad d: \begin{bmatrix} 0 & 0 & 0 & 1 \\ 1 & 0 & 0 & 0 \\ 0 & 1 & 0 & 0 \\ 0 & 0 & 1 & 0 \end{bmatrix} \tag{7.6}$$

because of the coincident 1's in the leftmost column. However, it can combine with the identity and the other two c matrices and the other two d matrices which have no coincident 1's to give the four matrices:

$$\begin{bmatrix} a & b & c & 0 \\ b & a & 0 & c \\ c & 0 & a & b \\ 0 & c & b & a \end{bmatrix} \quad \begin{bmatrix} a & b & c & 0 \\ b & a & 0 & c \\ 0 & c & a & b \\ c & 0 & b & a \end{bmatrix} \tag{7.7}$$

$$\begin{bmatrix} a & b & 0 & d \\ b & a & d & 0 \\ d & 0 & a & b \\ 0 & d & b & a \end{bmatrix} \quad \begin{bmatrix} a & b & 0 & d \\ b & a & d & 0 \\ 0 & d & a & b \\ d & 0 & b & a \end{bmatrix} \tag{7.8}$$

We are left to fill in the missing variables. We get the three copies of C_4 presented above, (6.5) & (6.6), but we also have another group:

$$C_2 \times C_2 = \begin{bmatrix} a & b & c & d \\ b & a & d & c \\ c & d & a & b \\ d & c & b & a \end{bmatrix} \tag{7.9}$$

This, (7.9), is an adjoint representation of an order four group. We cannot get to this arrangement, (7.9), of variables by any swapping of variables or swapping of the rows and columns of any of the three adjoint representations of C_4 given above, (6.5) & (6.6); therefore, we have a differently structured group – a different order four group.

The 'new' group is called $C_2 \times C_2$. We have now constructed all the adjoint representations of all order four groups.

Each of the four groups discovered above is a subgroup of the set of all 4×4 permutation matrices which is the symmetric group S_4. We can now say that S_4 has three C_4 subgroups and one $C_2 \times C_2$ subgroup and has no more order four subgroups.

Aside: The order five groups:
Of the 120 5×5 permutation matrices; there are:

a) 45 matrices with one 1 on the leading diagonal
b) 20 matrices with two 1s on the leading diagonal
c) 10 matrices with three 1s on the leading diagonal
d) There is one 5×5 matrix, the identity, with five 1's on the leading diagonal.
e) There cannot be any 5×5 matrices with four 1's on the leading diagonal for then there would be only one position for the remaining 1 and this would be on the leading diagonal.

It is a matter of combinatorial mathematics to calculate the number of $n \times n$ permutation matrices other than the identity with 1's on the leading diagonal.

We are now down from 120 prospective permutation matrices with which to form order five groups to only 44.

There are some 5×5 permutation matrices which, when raised to various powers, produce permutation matrices which have 1's on the leading diagonal but which are not the identity matrix. In the case of the 5×5 permutation matrices, these are all matrices whose second power produces the unwanted matrix; there are twenty such matrices; we can eliminate those matrices. We are now down to 24 prospective matrices plus the identity.

It is not always the case that raising matrices to only power two gives such unwanted matrices with unwanted 1's on the leading diagonal. In the case of the 7×7 permutation matrices, there are 714 such matrices at power 2, 630 such matrices at power 3, 1134 such matrices at power 4, 504 such matrices at power 5, and 924 such matrices at power 6.[23]

Summary:
We have succeeded in finding all the order four groups, and we have discovered how many of each type of order four group the order twenty-four symmetric group, S_4, has as subgroups.

However, above, we found that there are 24 5×5 permutation matrices with all zeros on the leading diagonal and whose powers have all zeros on the leading diagonal. Clearly, finding groups by casting out the matrices as we have done for the order four groups is a most laborious task for the higher order groups.

Repeated warning:
It is important to realise that not every set of permutation matrices which sum to an 'all 1's' matrix form a finite group. Consider:

$$\begin{bmatrix} 1 & 0 & 0 \\ 0 & 0 & 1 \\ 0 & 1 & 0 \end{bmatrix} + \begin{bmatrix} 0 & 0 & 1 \\ 0 & 1 & 0 \\ 1 & 0 & 0 \end{bmatrix} + \begin{bmatrix} 0 & 1 & 0 \\ 1 & 0 & 0 \\ 0 & 0 & 1 \end{bmatrix} \qquad (7.10)$$

These three permutation matrices, (7.10), do not form a finite group – for a start, there's no identity.

---

[23] Where would we be without computers?

Chapter 8

Adjoint Representations and Cyclic Groups

We have:

$$P_C = \begin{bmatrix} 0 & 1 & 0 & 0 \\ 0 & 0 & 1 & 0 \\ 0 & 0 & 0 & 1 \\ 1 & 0 & 0 & 0 \end{bmatrix} \qquad (8.1)$$

This matrix has 1's running immediately parallel to the leading diagonal and the other 'missing' 1 is in the bottom left-hand corner. For any size of matrix, we have a permutation matrix of this same basic form:

$$\begin{bmatrix} 0 & 1 & 0 \\ 0 & 0 & 1 \\ 1 & 0 & 0 \end{bmatrix} \& \begin{bmatrix} 0 & 1 & 0 & 0 & 0 \\ 0 & 0 & 1 & 0 & 0 \\ 0 & 0 & 0 & 1 & 0 \\ 0 & 0 & 0 & 0 & 1 \\ 1 & 0 & 0 & 0 & 0 \end{bmatrix} \& \begin{bmatrix} 0 & 1 & 0 & 0 & 0 & 0 \\ 0 & 0 & 1 & 0 & 0 & 0 \\ 0 & 0 & 0 & 1 & 0 & 0 \\ 0 & 0 & 0 & 0 & 1 & 0 \\ 0 & 0 & 0 & 0 & 0 & 1 \\ 1 & 0 & 0 & 0 & 0 & 0 \end{bmatrix}$$

$$(8.2)$$

We have:

$$\begin{bmatrix} 0 & 1 & 0 & 0 \\ 0 & 0 & 1 & 0 \\ 0 & 0 & 0 & 1 \\ 1 & 0 & 0 & 0 \end{bmatrix}^2 = \begin{bmatrix} 0 & 0 & 1 & 0 \\ 0 & 0 & 0 & 1 \\ 1 & 0 & 0 & 0 \\ 0 & 1 & 0 & 0 \end{bmatrix} \qquad (8.3)$$

$$\begin{bmatrix} 0 & 1 & 0 & 0 \\ 0 & 0 & 1 & 0 \\ 0 & 0 & 0 & 1 \\ 1 & 0 & 0 & 0 \end{bmatrix}^3 = \begin{bmatrix} 0 & 0 & 0 & 1 \\ 1 & 0 & 0 & 0 \\ 0 & 1 & 0 & 0 \\ 0 & 0 & 1 & 0 \end{bmatrix} \quad \begin{bmatrix} 0 & 1 & 0 & 0 \\ 0 & 0 & 1 & 0 \\ 0 & 0 & 0 & 1 \\ 1 & 0 & 0 & 0 \end{bmatrix}^4 = \begin{bmatrix} 1 & 0 & 0 & 0 \\ 0 & 1 & 0 & 0 \\ 0 & 0 & 1 & 0 \\ 0 & 0 & 0 & 1 \end{bmatrix}$$

(8.4)

Notice how the diagonal 1's 'dance' across the matrix as the power is increased. If the reader tries calculating the powers of the matrix by hand, the reader will understand that the powers of such a matrix will always 'dance' one step per power across the matrix.

Such matrices are called circulant matrices, and they have many applications[24].

Putting the powers, (8.1) & (8.3) & (8.4) together, we get:

$$C_4 = \begin{bmatrix} a & b & c & d \\ d & a & b & c \\ c & d & a & b \\ b & c & d & a \end{bmatrix} \quad (8.5)$$

This, (8.5), is a copy of the adjoint representation of the order four cyclic group C_4, (6.5).

Cyclic groups:
For any size of matrix, there is always a matrix of the form (8.1). It is a simple fact of matrix multiplication that such 'dancing' matrices always form a cyclic group of the appropriate order generated by the P_C type of matrix.

Thus, there is a cyclic group of every order $\{1, 2, 3, 4, 5, 6, 7, 8, 9, ...\}$. This is a standard theorem of finite group theory, but we have deduced and proved it by an unconventional route.

---

[24] See Davis: Circulant Matrices.

Adjoint Representations and Cyclic Groups

Thus, if we want a cyclic group of order n we simply write out the permutation matrices that 'run parallel' to the leading diagonal as $n \times n$ permutation matrices:

$$C_3 = \begin{bmatrix} a & b & c \\ c & a & b \\ b & c & a \end{bmatrix} \quad C_5 = \begin{bmatrix} a & b & c & d & e \\ e & a & b & c & d \\ d & e & a & b & c \\ c & d & e & a & b \\ b & c & d & e & a \end{bmatrix} \quad (8.6)$$

We call this the standard adjoint representation of the cyclic groups.

Even and odd permutations and the cyclic groups:
With thought, bearing in mind that swapping two columns swaps the sign of the determinant, we see that every 'dancing' matrix in the standard adjoint representation of a cyclic group with odd order, $\{C_1, C_3, C_5, ...\}$, is an even permutation – every one of the permutations in this, (8.6), is an even permutation.

The 'dancing' matrices in the standard adjoint representation of cyclic groups of even order, $\{C_2, C_4, ...\}$, are, in the order beginning with the b matrix, alternately odd and even permutation matrices.

The numbers of odd and even permutation matrices (permutations) is constant for different adjoint representations of the same group. It is always the case that an adjoint representation of a cyclic group of odd order is comprised of only even permutations. It is always the case that, when we include the identity, an adjoint representation of a cyclic group of even order is comprised of permutations half of which are odd and half of which are even. Remember, oddness and evenness is, in general, dependent upon the representation, circa (5.9), but oddness and evenness is invariant over different adjoint representations.

Aside:

The 'parallel to the leading diagonal' matrices are not the only types of permutation matrices which form adjoint representations of cyclic groups. For example, we saw above that there are two other ways of forming the adjoint representation of the order four cyclic group C_4 using 4×4 permutation matrices:

$$C_4 = \begin{bmatrix} a & b & c & d \\ b & a & d & c \\ d & c & a & b \\ c & d & b & a \end{bmatrix} \quad \& \quad C_4 = \begin{bmatrix} a & b & c & d \\ c & a & d & b \\ b & d & a & c \\ d & c & b & a \end{bmatrix} \quad (8.7)$$

Of course, these can be found by swapping variables and swapping columns, but finding them is neither efficient nor an easy task.

The adjoint cyclic group:

The adjoint representation of the cyclic group C_n is $n \times n$ permutation matrices.

For cyclic groups, the fundamental representation[25] is the same as the adjoint representation.

---

[25] Remember, the fundamental representation is the representation of smallest size permutation matrices that can accommodate the group as permutation matrices.

Chapter 9

Cosets

In this chapter, we digress a little as we delve deeper into finite group theory. We discover an easy way, using the adjoint representation of a group, to calculate the cosets of a group. Such a calculation would be quite laborious by conventional methods.

Cosets:
There are six 3×3 permutation matrices. They are the symmetric group S_3. Of these six permutation matrices, three form the order three cyclic group C_3.

$$C_3 = \left\{ \begin{bmatrix} 1 & 0 & 0 \\ 0 & 1 & 0 \\ 0 & 0 & 1 \end{bmatrix}, \begin{bmatrix} 0 & 1 & 0 \\ 0 & 0 & 1 \\ 1 & 0 & 0 \end{bmatrix}, \begin{bmatrix} 0 & 0 & 1 \\ 1 & 0 & 0 \\ 0 & 1 & 0 \end{bmatrix} \right\} \quad (9.1)$$

We are going to multiply each of these three matrices by each of the six 3×3 permutation matrices. First, we are going to multiply from the left.

If we multiply each of these three matrices by the identity, we simply get the same three matrices; we write:

$$\begin{bmatrix} 1 & 0 & 0 \\ 0 & 1 & 0 \\ 0 & 0 & 1 \end{bmatrix} C_3 = C_3 \quad (9.2)$$

Similarly, we get the group C_3 if we multiply these three matrices by either of the other two matrices in the group (9.1); we write:

$$\begin{bmatrix} 0 & 1 & 0 \\ 0 & 0 & 1 \\ 1 & 0 & 0 \end{bmatrix} C_3 = C_3 \qquad \begin{bmatrix} 0 & 0 & 1 \\ 1 & 0 & 0 \\ 0 & 1 & 0 \end{bmatrix} C_3 = C_3 \qquad (9.3)$$

Left cosets:

However, we have now used only three of the six 3×3 matrices. Suppose we multiply each of the three matrices in the group C_3 on the left by the other three 3×3 matrices. We get:

$$\begin{bmatrix} 1 & 0 & 0 \\ 0 & 0 & 1 \\ 0 & 1 & 0 \end{bmatrix} C_3 = \left\{ \begin{bmatrix} 1 & 0 & 0 \\ 0 & 0 & 1 \\ 0 & 1 & 0 \end{bmatrix}, \begin{bmatrix} 0 & 1 & 0 \\ 1 & 0 & 0 \\ 0 & 0 & 1 \end{bmatrix}, \begin{bmatrix} 0 & 0 & 1 \\ 0 & 1 & 0 \\ 1 & 0 & 0 \end{bmatrix} \right\} \qquad (9.4)$$

$$\begin{bmatrix} 0 & 1 & 0 \\ 1 & 0 & 0 \\ 0 & 0 & 1 \end{bmatrix} C_3 = \left\{ \begin{bmatrix} 0 & 1 & 0 \\ 1 & 0 & 0 \\ 0 & 0 & 1 \end{bmatrix}, \begin{bmatrix} 0 & 0 & 1 \\ 0 & 1 & 0 \\ 1 & 0 & 0 \end{bmatrix}, \begin{bmatrix} 1 & 0 & 0 \\ 0 & 0 & 1 \\ 0 & 1 & 0 \end{bmatrix} \right\} \qquad (9.5)$$

$$\begin{bmatrix} 0 & 0 & 1 \\ 0 & 1 & 0 \\ 1 & 0 & 0 \end{bmatrix} C_3 = \left\{ \begin{bmatrix} 0 & 0 & 1 \\ 0 & 1 & 0 \\ 1 & 0 & 0 \end{bmatrix}, \begin{bmatrix} 1 & 0 & 0 \\ 0 & 0 & 1 \\ 0 & 1 & 0 \end{bmatrix}, \begin{bmatrix} 0 & 1 & 0 \\ 1 & 0 & 0 \\ 0 & 0 & 1 \end{bmatrix} \right\} \qquad (9.6)$$

We see that we get the same set of three matrices, (9.4), (9.5), & (9.6). Bearing in mind that we multiplied from the left, we say that the three matrices in (9.4), (9.5), & (9.6) are a left coset of C_3 in S_3.

To generalise, we need to also say that the three matrices which are C_3 are also a left coset of C_3 in S_3. We say that the group C_3 has two left cosets in S_3. It is important to state both the group and the subgroup when speaking of cosets.

Right cosets:
Suppose that, instead of multiplying on the left, we multiply on the right. In this case, we get the same sets of matrices. We have:

$$\begin{bmatrix} 1 & 0 & 0 \\ 0 & 0 & 1 \\ 0 & 1 & 0 \end{bmatrix} C_3 = C_3 \begin{bmatrix} 1 & 0 & 0 \\ 0 & 0 & 1 \\ 0 & 1 & 0 \end{bmatrix} \qquad (9.7)$$

$$\begin{bmatrix} 0 & 1 & 0 \\ 1 & 0 & 0 \\ 0 & 0 & 1 \end{bmatrix} C_3 = C_3 \begin{bmatrix} 0 & 1 & 0 \\ 1 & 0 & 0 \\ 0 & 0 & 1 \end{bmatrix} \qquad (9.8)$$

$$\begin{bmatrix} 0 & 0 & 1 \\ 0 & 1 & 0 \\ 1 & 0 & 0 \end{bmatrix} C_3 = C_3 \begin{bmatrix} 0 & 0 & 1 \\ 0 & 1 & 0 \\ 1 & 0 & 0 \end{bmatrix} \qquad (9.9)$$

We see that the right cosets of C_3 in S_3 are equal to the left cosets of C_3 in S_3. This does not happen with every subgroup of a group. When the left cosets of a subgroup are the same as the right cosets of a subgroup, we say the subgroup is a normal subgroup of the larger group.

Cosets in abelian groups:
Clearly, right cosets are equal to left cosets in an abelian[26] group. Thus all subgroups of an abelian group are normal.

Making things simpler:
Let us write the group C_3 as:

---

[26] Abelian groups are groups in which every pair of elements commutes.

$$C_3 = \begin{bmatrix} a & b & c \\ c & a & b \\ b & c & a \end{bmatrix} \qquad (9.10)$$

Let us now multiply this group from the left by a matrix which is not one of the group matrices as we did above in (9.4)

$$\begin{bmatrix} 1 & 0 & 0 \\ 0 & 0 & 1 \\ 0 & 1 & 0 \end{bmatrix} \begin{bmatrix} a & b & c \\ c & a & b \\ b & c & a \end{bmatrix} = \begin{bmatrix} a & b & c \\ b & c & a \\ c & a & b \end{bmatrix} \qquad (9.11)$$

The product on the right of (9.11) is equivalent to:

$$\left\{ \begin{bmatrix} 1 & 0 & 0 \\ 0 & 0 & 1 \\ 0 & 1 & 0 \end{bmatrix}, \begin{bmatrix} 0 & 1 & 0 \\ 1 & 0 & 0 \\ 0 & 0 & 1 \end{bmatrix}, \begin{bmatrix} 0 & 0 & 1 \\ 0 & 1 & 0 \\ 1 & 0 & 0 \end{bmatrix} \right\} \qquad (9.12)$$

Of course, a permutation matrix permutes the rows of the matrix it acts upon. We see that we can calculate the cosets of a subgroup using the Standard Form Cayley table of the group.

This is one of the insights to group theory that working with permutation matrices brings. How much easier it is to calculate cosets.

How about doing it all at once:

$$\begin{bmatrix} x & y & z \\ y & z & x \\ z & x & y \end{bmatrix} \begin{bmatrix} a & b & c \\ c & a & b \\ b & c & a \end{bmatrix} = \begin{bmatrix} ax+bz+cy & ay+bx+cz & az+by+cx \\ ay+bx+cz & az+by+cx & ax+bz+cy \\ az+by+cx & ax+bz+cy & ay+bx+cz \end{bmatrix} =$$

$$ax+bz+cy \begin{bmatrix} 1 & 0 & 0 \\ 0 & 0 & 1 \\ 0 & 1 & 0 \end{bmatrix} + ay+bx+cz \begin{bmatrix} 0 & 1 & 0 \\ 1 & 0 & 0 \\ 0 & 0 & 1 \end{bmatrix} + az+by+cx \begin{bmatrix} 0 & 0 & 1 \\ 0 & 1 & 0 \\ 1 & 0 & 0 \end{bmatrix}$$

(9.13)

Cosets

Using a different subgroup:

Suppose, instead of using the C_3 subgroup of S_3, in its stead, we use a C_2 subgroup. We choose the C_2 subgroup[27]:

$$C_2' = \left\{ \begin{bmatrix} 1 & 0 & 0 \\ 0 & 1 & 0 \\ 0 & 0 & 1 \end{bmatrix}, \begin{bmatrix} 0 & 0 & 1 \\ 0 & 1 & 0 \\ 1 & 0 & 0 \end{bmatrix} \right\} \qquad (9.14)$$

We have put the prime on the C_2 to distinguish it from the other two C_2 subgroups in S_3. The left and right cosets of the b variable are:

$$\begin{bmatrix} 0 & 1 & 0 \\ 0 & 0 & 1 \\ 1 & 0 & 0 \end{bmatrix} C_2' = \left\{ \begin{bmatrix} 0 & 1 & 0 \\ 0 & 0 & 1 \\ 1 & 0 & 0 \end{bmatrix}, \begin{bmatrix} 0 & 1 & 0 \\ 1 & 0 & 0 \\ 0 & 0 & 1 \end{bmatrix} \right\} \qquad (9.15)$$

$$C_2' \begin{bmatrix} 0 & 1 & 0 \\ 0 & 0 & 1 \\ 1 & 0 & 0 \end{bmatrix} = \left\{ \begin{bmatrix} 0 & 1 & 0 \\ 0 & 0 & 1 \\ 1 & 0 & 0 \end{bmatrix}, \begin{bmatrix} 1 & 0 & 0 \\ 0 & 0 & 1 \\ 0 & 1 & 0 \end{bmatrix} \right\} \qquad (9.16)$$

We see that the left coset is not the same as the right coset. The subgroup C_2', (9.14), is not a normal subgroup of S_3.

Taking:

$$C_2' = \left\{ \begin{bmatrix} 1 & 0 & 0 \\ 0 & 1 & 0 \\ 0 & 0 & 1 \end{bmatrix}, \begin{bmatrix} 0 & 0 & 1 \\ 0 & 1 & 0 \\ 1 & 0 & 0 \end{bmatrix} \right\} = \begin{bmatrix} a & 0 & c \\ 0 & a+c & 0 \\ c & 0 & a \end{bmatrix} \qquad (9.17)$$

And multiplying to the left or to the right as in (9.15) & (9.16) will give the same results – Hey! It works all the time. This really is an easier way of calculating cosets.

---

[27] This is neither the adjoint representation nor the fundamental representation of the C_2 group.

Aside:

We could express the above (9.15) & (9.16) using the adjoint representation of S_3 as:

$$\begin{bmatrix} 0 & 1 & 0 & 0 & 0 & 0 \\ 0 & 0 & 1 & 0 & 0 & 0 \\ 1 & 0 & 0 & 0 & 0 & 0 \\ 0 & 0 & 0 & 0 & 1 & 0 \\ 0 & 0 & 0 & 0 & 0 & 1 \\ 0 & 0 & 0 & 1 & 0 & 0 \end{bmatrix} \begin{bmatrix} a & 0 & 0 & 0 & 0 & f \\ 0 & a & 0 & 0 & f & 0 \\ 0 & 0 & a & f & 0 & 0 \\ 0 & 0 & f & a & 0 & 0 \\ 0 & f & 0 & 0 & a & 0 \\ f & 0 & 0 & 0 & 0 & a \end{bmatrix} = \begin{bmatrix} 0 & a & 0 & 0 & f & 0 \\ 0 & 0 & a & f & 0 & 0 \\ a & 0 & 0 & 0 & 0 & f \\ 0 & f & 0 & 0 & a & 0 \\ f & 0 & 0 & 0 & 0 & a \\ 0 & 0 & f & a & 0 & 0 \end{bmatrix}$$

(9.18)

$$\begin{bmatrix} a & 0 & 0 & 0 & 0 & f \\ 0 & a & 0 & 0 & f & 0 \\ 0 & 0 & a & f & 0 & 0 \\ 0 & 0 & f & a & 0 & 0 \\ 0 & f & 0 & 0 & a & 0 \\ f & 0 & 0 & 0 & 0 & a \end{bmatrix} \begin{bmatrix} 0 & 1 & 0 & 0 & 0 & 0 \\ 0 & 0 & 1 & 0 & 0 & 0 \\ 1 & 0 & 0 & 0 & 0 & 0 \\ 0 & 0 & 0 & 0 & 1 & 0 \\ 0 & 0 & 0 & 0 & 0 & 1 \\ 0 & 0 & 0 & 1 & 0 & 0 \end{bmatrix} = \begin{bmatrix} 0 & a & 0 & f & 0 & 0 \\ 0 & 0 & a & 0 & 0 & f \\ a & 0 & 0 & 0 & f & 0 \\ f & 0 & 0 & 0 & a & 0 \\ 0 & 0 & f & 0 & 0 & a \\ 0 & f & 0 & a & 0 & 0 \end{bmatrix}$$

(9.19)

Summary:

In this chapter, we have done a little finite group theory using permutation matrices. We have seen that using permutation matrices will sometimes provide insights into group theory which are not obvious using standard group theory techniques. To continue in this direction would bring more insights but would be to repeat work done in standard group theory texts. This is a worth-while endeavour, but, other than the example in this chapter, we do not walk that road in this book.

We have, at least, found an easy way to calculate cosets.

We continue in the next chapter with a whole now vista upon group theory.

Chapter 10

Standard Adjoints

Above, circa (8.1) onward, we saw that we are able to write the adjoint representation of any cyclic group as variables running parallel to the leading diagonal, (8.6). There are many other kinds of finite groups, dihedral groups, symmetric groups, alternating groups, dicyclic groups, the sporadic groups etc.. Is there a standard form of adjoint representation for each kind of group?

This question is unanswered at present, but if the answer were to become known and if the answer is yes, then, using permutation matrices and standard adjoint representations, perhaps we could construct a complete catalogue of all finite groups.

Aside:
Mathematics does not yet have a complete catalogue of all finite groups, but we do have a catalogue of all simple finite groups. It took mathematicians 150 years to compile this catalogue of simple finite groups.

The standard adjoint representations of dihedral groups:
There is a known standard adjoint representation of every dihedral group. This particular adjoint representation is of the form:

$$\begin{bmatrix} [C_n]_{Unflipped} & [C_n]_{Flipped} \\ [C_n]_{Flipped} & [C_n]_{Unflipped} \end{bmatrix} \quad (10.1)$$

We will use C_3 to illustrate this. The adjoint representation of the order three finite cyclic group C_3 is:

$$C_3 = \begin{bmatrix} a & b & c \\ c & a & b \\ b & c & a \end{bmatrix} \qquad (10.2)$$

We have, see (10.1):

$$D_3 \sim \begin{bmatrix} \begin{bmatrix} a & b & c \\ c & a & b \\ b & c & a \end{bmatrix} & [\ ] \\ [\ ] & \begin{bmatrix} a & b & c \\ c & a & b \\ b & c & a \end{bmatrix} \end{bmatrix} = \begin{bmatrix} \begin{matrix} a & b & c \\ c & a & b \\ b & c & a \end{matrix} & \\ & \begin{matrix} a & b & c \\ c & a & b \\ b & c & a \end{matrix} \end{bmatrix} \qquad (10.3)$$

We then 'flip' the adjoint representation of the appropriate cyclic group. We change the variables to avoid confusion and because the dihedral group is of order twice the order of the appropriate cyclic group:

$$\begin{bmatrix} a & b & c \\ c & a & b \\ b & c & a \end{bmatrix} \xrightarrow{Flip} \begin{bmatrix} c & b & a \\ b & a & c \\ a & c & b \end{bmatrix} \xrightarrow{Re-name} \begin{bmatrix} d & e & f \\ e & f & d \\ f & d & e \end{bmatrix} \qquad (10.4)$$

We then place this matrix in the two blank places in (10.3) to produce:

$$D_3 = \begin{bmatrix} a & b & c & d & e & f \\ c & a & b & e & f & d \\ b & c & a & f & d & e \\ d & e & f & a & b & c \\ e & f & d & c & a & b \\ f & d & e & b & c & a \end{bmatrix} \qquad (10.5)$$

This, (10.5), is the standard adjoint representation of the dihedral group D_3.

Standard Adjoints

The standard adjoint representation of D₄:

Similarly, using the Standard Form Cayley table of the C_4 group, we can construct the Standard adjoint representation of the order eight dihedral group, D_4:

$$D_4 = \begin{bmatrix} a & b & c & d & e & f & g & h \\ d & a & b & c & f & g & h & e \\ c & d & a & b & g & h & e & f \\ b & c & d & a & h & e & f & g \\ e & f & g & h & a & b & c & d \\ f & g & h & e & d & a & b & c \\ g & h & e & f & c & d & a & b \\ h & e & f & g & b & c & d & a \end{bmatrix} \quad (10.6)$$

All the dihedral groups have standard adjoint representations that can be similarly constructed.

Oddness and evenness:

Consider an odd 3×3 permutation matrix like:

$$\begin{bmatrix} 1 & 0 & 0 \\ 0 & 0 & 1 \\ 0 & 1 & 0 \end{bmatrix} \quad (10.7)$$

Place this permutation matrix upon the leading diagonal into a matrix of twice its size:

$$\begin{bmatrix} 1 & 0 & 0 & 0 & 0 & 0 \\ 0 & 0 & 1 & 0 & 0 & 0 \\ 0 & 1 & 0 & 0 & 0 & 0 \\ 0 & 0 & 0 & 1 & 0 & 0 \\ 0 & 0 & 0 & 0 & 0 & 1 \\ 0 & 0 & 0 & 0 & 1 & 0 \end{bmatrix} \quad (10.8)$$

We now have a 6×6 even permutation matrix. Similarly, if we place an even permutation matrix upon the leading diagonal into a matrix of twice its size, we also get an even permutation matrix.

We see that the permutations within the standard adjoint representation of the dihedral group, D_n, which occupy the $\frac{n}{2}$ leading diagonal blocks are always even permutations.

In the case of D_3, (10.5), the permutation matrices associated with the $\{a,b,c\}$ variables are all even permutations, and the permutation matrices associated with the $\{d,e,f\}$ variables are all odd permutations.

In the case of D_4, (10.6), the permutation matrices associated with the $\{a,b,c,d\}$ variables are all even permutations, and the permutation matrices associated with the $\{e,f,g,h\}$ variables are also all even permutations.

When the subscript of the D, which is half the order of the dihedral group, is odd, the permutation matrices corresponding to the variables in the right-most half of the top row of the adjoint representation of the dihedral group are all odd. When the subscript of the D, which is half the order of the dihedral group, is even, the permutation matrices corresponding to the variables in the right-most half of the top row of the adjoint representation of the dihedral group are all even. This is a general phenomenon of dihedral groups.

Dihedral groups with an odd subscript are comprised of half even permutations and half odd permutations.

Dihedral groups with an even subscript are comprised of all even permutations.

Standard adjoint representations of symmetric groups:

The dihedral group D_3 is also the symmetric group S_3 - two names, same thing. We might be tempted to assume the standard adjoint representation of symmetric groups is of the same form as the standard adjoint of S_3. The standard adjoint representation of the symmetric groups is not yet known.

Summary:

We have seen that there are standard adjoint representations of some types of finite group. We conjecture that the same is true of all types of finite group. There is research to be done here by anyone with a computer and a mathematics software package[28].

---

[28] Your author would be most grateful to be kept informed of any developments in this area – his address is in the front of this book.

Chapter 11

The Important Stuff

So far in this book, we have played around with permutation matrices in a childish way. Permutation matrices are simple, but we have found that they lead us directly into finite group theory. As we have played around with these permutation matrices, we have discovered some surprising insights into finite group theory, but nothing spectacular. We might say that permutation matrices are an amusing backwater of mathematics which might make a pleasant Christmas present for young children. Remarkably, these simple little matrices are the foundations of all mathematics and of modern theoretical physics.

Perhaps we should say finite groups are the foundations of all mathematics and of modern theoretical physics. No, we should not say this; it is the permutation matrices that go directly to modern physics and mathematics. All we need add to the permutation matrices is the concept of multiplying them by a real number and the concept of the exponential function acting upon a matrix.

The special theory of relativity:
We have:

$$\begin{bmatrix} \chi & 0 \\ 0 & \chi \end{bmatrix} \begin{bmatrix} 0 & 1 \\ 1 & 0 \end{bmatrix} = \begin{bmatrix} 0 & \chi \\ \chi & 0 \end{bmatrix}$$

$$\exp\left(\begin{bmatrix} 0 & \chi \\ \chi & 0 \end{bmatrix}\right) = \begin{bmatrix} \cosh \chi & \sinh \chi \\ \sinh \chi & \cosh \varphi \end{bmatrix} = \begin{bmatrix} \gamma & v\gamma \\ v\gamma & \gamma \end{bmatrix} \quad (11.1)$$

$$\gamma = \frac{1}{\sqrt{1 - \dfrac{v^2}{c^2}}}$$

The matrix with the hyperbolic trigonometric functions as elements is the rotation matrix of 2-dimensional space-time known as the 2-dimensional Lorentz transformation[29]. Rotation in 2-dimensional space-time is change of velocity. The special theory of relativity is no more than the statement that "physics is invariant under rotation in 2-dimensional space-time"; this is, "physics is invariant under change of velocity".

The physics of, say, a car engine or reacting chemicals in a test-tube do not change when we point the engine or the test-tube north rather than west. So it is with rotation in space-time; rotation in 2-dimensional space-time does not change the physics; the statement of this fact is the special theory of relativity.

The rotation matrix in 2-dimensional space-time is no more than the exponential of a 2×2 permutation matrix multiplied by a real number (the angle).

The complex numbers:

Let us throw a minus sign into the odd 2×2 permutation matrix and again multiply both 2×2 permutation matrices by real numbers:

$$\begin{bmatrix} a & b \\ -b & a \end{bmatrix} = \mathbb{C}$$

$$\exp\left(\begin{bmatrix} a & b \\ -b & a \end{bmatrix}\right) = \begin{bmatrix} r & 0 \\ 0 & r \end{bmatrix} \begin{bmatrix} \cos b & \sin b \\ -\sin b & \cos b \end{bmatrix}$$

(11.2)

We have the complex numbers. We have included the polar form of the complex numbers. This is just the exponential of permutation matrices.

---

[29] The phrase 'Lorentz transformation' is used ambiguously in physics, and so we append the '2-dimensional' words to avoid confusion.

The quaternions:

Consider the adjoint representation of the commutative finite group $C_2 \times C_2$, (7.9), and throw some minus signs into it:

$$C_2 \times C_2 = \begin{bmatrix} a & b & c & d \\ b & a & d & c \\ c & d & a & b \\ d & c & b & a \end{bmatrix} \rightarrow \begin{bmatrix} a & b & c & d \\ -b & a & -d & c \\ -c & d & a & -b \\ -d & -c & b & a \end{bmatrix} = \mathbb{H}_{Lx} \quad (11.3)$$

We have the quaternions. There are similar division algebras to be derived from every adjoint representation of any finite group[30].

Gauge theory:

We will differentiate a complex number with respect to two spatial variables, $\{x, y\}$. We cannot differentiate outside of a division algebra, and so we form these two variables into a complex number, but we take the view that they are spatial variables in our 4-dimensional space-time. We are going to differentiate to discover how a complex number varies as it moves over our 4-dimensional space-time. We have[31]:

$$\frac{\partial \begin{bmatrix} f(x,y) & g(x,y) \\ -g(x,y) & f(x,y) \end{bmatrix}}{\partial \begin{bmatrix} x & y \\ -y & x \end{bmatrix}} = \begin{bmatrix} \frac{\partial f}{\partial x} + \frac{\partial g}{\partial y} & \frac{\partial g}{\partial x} - \frac{\partial f}{\partial y} \\ -\left(\frac{\partial g}{\partial x} - \frac{\partial f}{\partial y}\right) & \frac{\partial f}{\partial x} + \frac{\partial g}{\partial y} \end{bmatrix}$$

$$= \begin{bmatrix} \text{Div} & \text{Curl} \\ -\text{Curl} & \text{Div} \end{bmatrix} \quad (11.4)$$

The divergence corresponds to a quantity which can be conserved like mass or electric charge. The curl is a force. We have a force which varies locally over our 4-dimensional space-time. If we were to write the complex number field in polar form, this locally varying force

---

[30] See: Dennis Morris : Quaternions.
[31] See: Dennis Morris : Complex Numbers The Higher Dimensional Forms.

would correspond to a locally varying angle, phase, over our 4-dimensional space-time. This is $U(1)$ gauge theory. $U(1)$ gauge theory is central to particle physics.

If we began the differentiation with a quaternion instead of a complex number and differentiated non-commutatively[32], we would have $SU(2)$ gauge theory[33]. $SU(2)$ gauge theory is central to the electro-weak unification of particle physics.

We see that our childish permutation matrices have led us directly to the heart of particle physics.

The general theory of relativity:
Although we do not show it here, the A_3 permutation matrices associated with the order four $C_2 \times C_2$ group lead to our 4-dimensional space-time and general relativity.

---

[32] See: Dennis Morris – The Physics of Empty Space.
[33] The commutation relations of the quaternions are $SU(2)$.

Chapter 12

Concluding Remarks

We began with the babyish operation of simply putting a single 1 into every row and every column of a square matrix. We were led to permutations and hence to finite group theory. We lingered a little in the finite groups and found insights which are not recognised by the more usual presentation of finite group theory. We left finite groups with much work still undone; perhaps the reader will choose this area of research for herself. We then discovered that permutation matrices underlie the foundational concept of mathematics which is number; we derived the complex numbers and the quaternions, but we also derived the less well known hyperbolic complex numbers which hold the special theory of relativity. We went on to construct gauge theory.

Although we have not shown it in this book, there are types of complex numbers of all dimension and they all arise directly from permutation matrices by taking the matrix exponential. Although we have not shown it in this book, huge areas of theoretical physics including general relativity, gauge theory and the fermion content of the standard model of particle physics derive directly from permutation matrices. Indeed, although the work is not yet complete, it seems that the whole of mathematics and the whole of our physical universe derives from permutation matrices. This is quite profound.

As we mentioned just above, there is much 'simple' work to be done and unknown truths to be discovered within permutation matrices. This work requires little more mathematics than elementary matrix algebra. There is also much work to be done in theoretical physics with permutation matrices.

Your author hopes this book has been an easy, pleasant, and interesting read. Your author hopes the reader has been inspired by the profound mathematics which derives from such simple objects as permutation matrices. Perhaps the reader will go on to do their own research in this area of mathematics. Perhaps, one Christmas day, the reader will introduce children to this area of mathematics and thereby inspire those children with a lifelong love of mathematics.

For now, we leave this subject incomplete and awaiting the attentions of others who will build upon the basic foundation presented in this book.

Thank you for your attention.

Dennis Morris

Brotton (May 2017)

Other Books by Dennis Morris

The Naked Spinor – A Rewrite of Clifford Algebra

Spinors exist in Clifford algebras. In this book, we explore the nature of spinors. This book is an excellent introduction to Clifford algebra.

Complex Numbers The Higher Dimensional Forms – Spinor Algebra

In this book, we explore the higher dimensional forms of complex numbers. These higher dimensional forms are connected very closely to spinors.

Upon General Relativity

In this book, we see how 4-dimensional space-time, gravity, and electromagnetism emerge from the spinor algebras. This is an excellent and easy-paced introduction to general relativity.

From Where Comes the Universe

This is a guide for the lay-person to the physics of empty space.

Empty Space is Amazing Stuff – The Special Theory of Relativity

This book deduces the theory of special relativity from the finite groups. It gives a unique insight into the nature of the 2-dimensional space-time of special relativity.

The Nuts and Bolts of Quantum Mechanics

This is a gentle introduction to quantum mechanics for undergraduates.

Quaternions

This book pulls together the often separate properties of the quaternions. Non-commutative differentiation is covered as is non-commutative rotation and non-commutative inner products along with the quaternion trigonometric functions.

The Uniqueness of our Space-time

This book reports the finding that the only two geometric spaces within the finite groups are the two spaces that together form our universe. This is a startling finding. The nature of geometric space is explained alongside the nature of division algebra space, spinor space. This book is a catalogue of the higher dimensional complex numbers up to dimension fifteen.

Lie Groups and Lie Algebras

This book presents Lie theory from a diametrically different perspective to the usual presentation. This makes the subject much more intuitively obvious and easier to learn. Included is perhaps the clearest and simplest presentation of the true nature of the Lie group $SU(2)$ ever presented.

The Physics of Empty Space

This book presents a comprehensive understanding of empty space. The presence of 2-dimensional rotations in our 4-dimensional space-time is explained. Also included is a very gentle introduction to non-commutative differentiation. Classical electromagetism is deduced from the quaternions.

The Electron

This book presents the quantum field theory view of the electron and the neutrino. This view is radically different from the classical view of the electron presented in most schools and colleges. This book gives a very clear exposition of the Dirac equation including the quaternion rewrite of the Dirac equation. This is an excellent introduction to particle physics for students prior to university, during university and after university courses in physics.

The Quaternion Dirac Equation

This small book (only 40 pages) presents the quaternion form of the Dirac equation. The neutrino mass problem is solved and we gain an explanation of why neutrinos are left-chiral. Much of the material in this book is drawn from 'The Electron'; this material is presented concisely and inexpensively for students already familiar with QFT.

An Essay on the Nature of Space-time

This small and inexpensive volume presents a view of the nature of empty space without the detailed mathematics. The expanding universe and dark energy is discussed.

Elementary Calculus from an Advanced Standpoint

This book rewrite the calculus of the complex numbers in a way that covers all division algebras and makes all continuous complex functions differentiable and integrable. Non-commutative differentiation is covered. Gauge covariant differentiation is covered as is the covariant derivative of general relativity.

Even Mathematicians and Physicists make Mistakes

This book points out what seems to be several important errors of modern physics and modern mathematics. Errors like the misunderstanding of rotation, the failure to teach the higher dimensional complex numbers in most universities, and the mathematical inconsistency of the Dirac equation and some casual errors are discussed. These errors are set in their historical circumstances and there is discussion about why they happened and the consequences of their happening. There is also an interesting chapter on the nature of mathematical proof within our society, and several famous proofs are discussed (without the details).

Other Books by Dennis Morris

Finite Groups – A Simple Introduction

This book introduces the reader to finite group theory. Many introductory books on finite groups bury the reader in geometrical examples or in other types of groups and lose the central nature of a finite group. This book sticks firmly with the permutation nature of finite groups and elucidates that nature by the extensive use of permutation matrices. Permutation matrices simplify the subject considerably. This book is probably unique in its use of permutation matrices and therefore unique in its simplicity.

The Left-handed Spinor

This book covers the left-handed parts of mathematics which we call the chiral algebras. These algebras have CP invariance, violation of parity, and many other aspects which makes them relevant to theoretical physics. It is quite a revelation to discover that mathematics is left-handed.

Index

2

2-dimensional space-time, 71

3

3 x 3 permutation matrices, 30
3-dimensional complex numbers, 45
3-dimensional rotation matrix, 46

A

abelian group, 61
adjoint representation, 43, 44
adjoint representation, C3, 45
adjoint representation, Cayley table, 48
adjoint representation, S3, 44
adjoint representations of cyclic groups, 55
all 1's' matrix, 48
alternating group, 33
alternating group, A3, 40
associativity, 32

C

C2 x C2 group, 52
C2 x C2, commutative finite group, 72
calculative procedures, 9
Cayley table, 46
Cayley, Arthur, 26
circulant matrices, 56
closure, permutation matrices, 9
closure, sequential combination, 6
coloured balls, 2
complex numbers, 71
cosets, 59
cosets, calculation of, 60
counter identity matrix, 28
cube root of minus unity, 27
curl, 72
cyclic group, 29
cyclic group of three permutation matrices, 29
cyclic group, C3, 38
cyclic group, C4, 47
cyclic group, C4, adjoint representations, 47
cyclic groups, adjoint representations, 55
cyclic groups, odd or even permutations, 57
cyclic groups, representations, 45

D

danceing matrices, 56
determinant, 15
determinant of a permutation matrix, 19
determinant, calculation of, 21
determinant, general properties of, 21
determinant, identity matrix, 20
determinant, nature of, 21
determinant, sign of, 19
determinant, swapping columns, 19
dicyclic groups, 65
differentiation, 72
dihedral groups, 65
dihedral groups, standard adjoint representation, 65
divergence, 72
division algebra, 9

E

electro-weak unification, 73
even permutation matrices, 24
exponential, 45, 71

Index

F

finite group, 32
finite group theory, 32
finite group, definition, 32
finite group, order 1, 34
finite group, order 2, 34
finite group, order 3, 36
fundamental representation, 43

G

gauge theory, 73
general relativity, 73
generating elements of a group, 30
genuine multiplication, 9

I

identity matrix, 6
identity permutation, 5
inverse of a permutation matrix, 25
inverse permutation, 10
inverse permutation, calculation of, 11

L

leading diagonal, 11
left cosets, 60
Levi-Civita symbol, 17
Levi-Civta symbol, 14
Lorentz transformation, 71

M

matrix determinant, 15
matrix multiplication, 7
matrix transposition, 11
matrix, power of, 27
matrix, quaternion elements, 22
multiplicative closure, 9

N

n-factorial, 2
non-commutivity, 12

normal subgroup, 61

O

odd permutation matrices, 24
oddness and evenness, adjoint representation of dihedral group, 68
oddness/evenness, representation, 38
one-to-one correspondence, 3
order, 33
order four groups, 50
order of a permutation matrix, 26, 27
order, of a finite group, 33

P

particle physics, 73
permutation, 2
permutation matrices 7 x 7, 54
permutation matrices, 5 x 5, 53
permutation matrices, closure, 9
permutation matrices, multiplication, 7
permutation matrices, number of, 12
permutation matrices, odd or even, 24
permutation matrix, definition, 2
permutation matrix, determinant, 19
permutation matrix, from a permutation, 4
permutation matrix, order of, 27
permutation of objects, 2
permutation, conventional notation, 3
permutation, identity, 5
permutation, inverse, 5
permutation, sequential combination of, 6
polar form of a division algebra, 45
polar form of the complex numbers, 71
power of matrices, 27
product of two permutation matrices, 9

Q

quaternions, 72

R

representation, oddness and evenness, 38
representations, 43
representations of finite groups, 34
reverse permutation, 10
right cosets, 61
roots of unity, 26
rotation matrix, 45

S

sequential combination of permutations, 7
sequential combination, closure, 6
sign of the determinant, 19
simple finite groups, 65
special theory of relativity, 70
sporadic groups, 65
square root of minus unity, 27
standard adjoint representation, 57
standard adjoint representation of the cyclic groups, 57
Standard Form Cayley table, 47
Standard Form Cayley table, coset calculation, 62
standard form of adjoint representation, 65
SU(2) gauge theory, 73
subgroup, 40
subgroups of symmetric groups, 40
symmetric group, 30
symmetric group, denotation, 31
symmetric group, S3, as 6 x 6 matrix, 44
symmetric groups, 39
symmetric permutation matrices, 28

T

transpose matrix, 25
transposition, 11
trigonometric functions, 45

U

U(1) gauge theory, 73

Z

zero divisors, 9

Printed in Great Britain
by Amazon

The Little Book
of
Permutation Matrices

By

Dennis Morris

Copyright: Dennis Morris

dennis355@btinternet.com

All Rights Reserved

Published by: Abane & Right

56 Coach Road

Brotton

Saltburn-by-the-Sea

TS12 2RP

01287 678918

May 2017